Making Construction Vehicles For Kids

By Luc St-Amour

Fox
Chapel Publishing Co. Inc.
1970 Broad Street • East Petersburg, PA 17520 • www.carvingworld.com

Publisher: Alan Giagnocavo
Editor: Ayleen Stellhorn
Desktop Specialist: Linda L. Eberly, Eberly Designs Inc.
Cover Photography: Robert Polett

ISBN # 1–56523–151–1
Library of Congress Card Number: 00–193338

To order your copy of this book,
please send check or money order
for the cover price plus $3.00 shipping to:
Fox Books
1970 Broad Street
East Petersburg, PA 17520

Or visit us on the web at
www.carvingworld.com

Manufactured in the USA

IMPORTANT NOTICE TO PARENTS AND READERS

The models in this book are recommended only for children five years of age and up. Models contain small parts that may break and be swallowed by a young child, posing a choking hazard.

- Always use non-toxic products to finish the models.
- Use water-proof glue if the models are to be used outside.
- Check models periodically to ensure that parts have not broken or become loose.
- When using string, be sure to use pieces that are short in length to avoid injuries.

Table of Contents

Introduction

This book is for all those who enjoy building wooden models. In this edition you will learn to build fascinating construction vehicles. Our main objective in creating this book was to help you attain the best possible results in reproducing these vehicles as easily and simply as possible. Building them require patience, know-how and, most importantly, the proper tools.

You will note that special care has been taken to provide you with as many explanations and instructions as possible. All patterns are full size and easily transfered onto wood. Easy-to-follow assembly drawings are also included for each model.

Please keep in mind the most important aspect of woodworking - safety!

I wish you the best of luck with your projects.

Please note: The models you can build with this book are suitable for children 5 years of age and up. If you wish to make display models with more parts and features, you can buy the book entitled *Realistic Construction Models You Can Make* by the same author.

Acknowledgments

I wish to thank all the people who have contributed to this project, especially my family and friends.

I would also like to thank the "Autodesk" company, who supplied me with the "Autocad" computer software.

I dedicate this book to all the people who are young at heart - people who still believe that dreams can come true and that everyday life is full of little joys.

Luc St-Amour

How to use this book

1. Start by reading through the book to get familiar with its contents.
2. Make the two jigs shown on page 4 and 5.
3. Use the materials list to cut all the parts required for a particuliar model. Label the parts with a pencil using their corresponding number (e.g. L1, L2)
4. You have been supplied with two sets of patterns for each model. The first set is found with the instructions and includes all parts. The second set can be found in the appendix and includes only those parts complicated enough to require a pattern. Using scissors, cut out the second set of patterns needed to make the model you have chosen to build.
5. Attach the pattern to the proper piece of stock.
6. Cut and sand the finished parts.
7. Mark drill holes, if required, and remove the pattern.
8. When all the parts are completed, follow the step-by-step assembly drawings to complete your model.

Recommended Tools

The following is a list of recommended tools. These power tools will give you the precision needed to make the models and will also save you a lot of time.

To make some parts you will be required to make inside cuts. The best tool for this task is the scroll saw.

For sanding purposes, you will need power sanders to save you time and give better results.

The wood needed to make these projects varies in thicknesses which are not standard. This means you will need to use a thickness planner and a bandsaw to re-saw and bring the wood to the specified thickness. (Some of you may have access to these tools at school, from friends or at a store.)

Accessories Needed

- Drill bit set (1/16" to 1/2")
- Brad point bit (1/8")
- Flat drill bit set (3/8" to 1")
- Measuring tape
- Combination square
- Sanding drum set
- Wood vice
- Assorted c-clamps
- Scriber
- Wood glue
- Wood file
- Pencil and eraser
- 12" ruler (clear recommended)
- 1/2" wide masking tape
- Compass

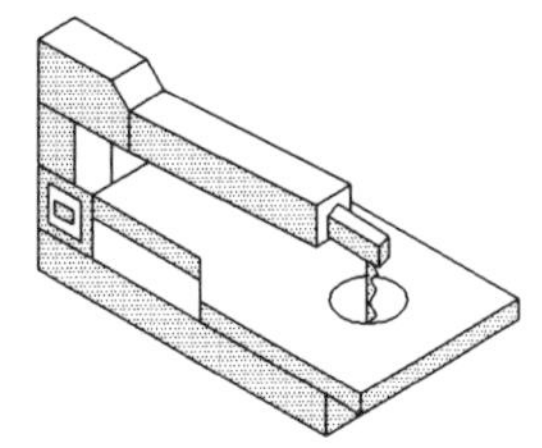

Scroll Saw

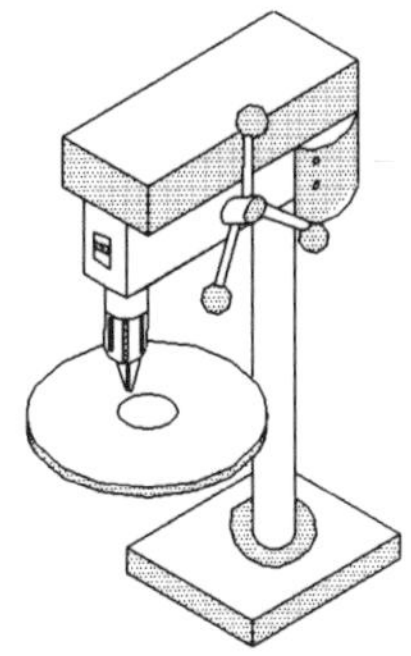

Drill Press

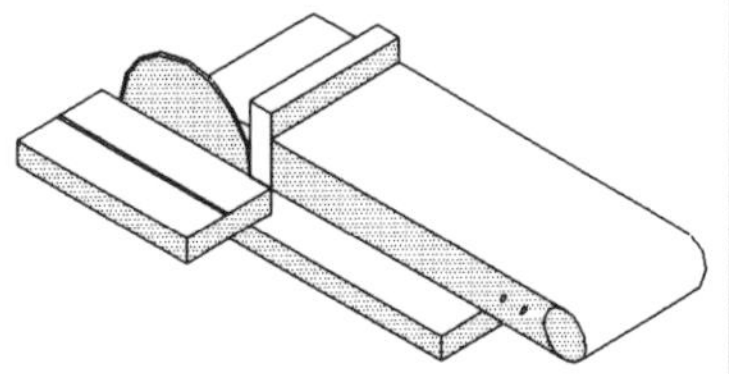

Belt/Disc Sander

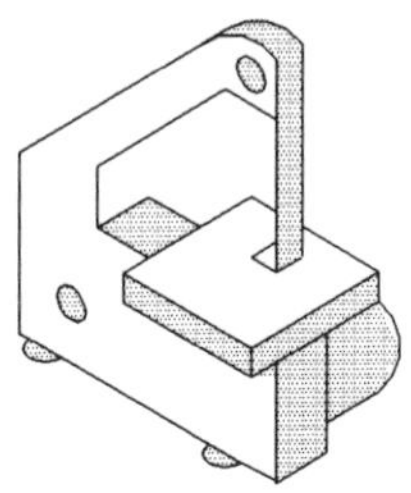

1" Belt Sander

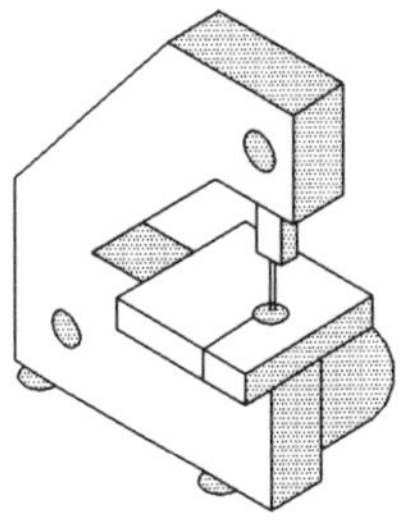

Band Saw

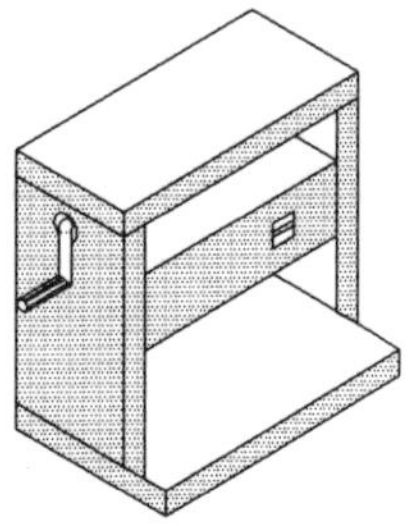

Thickness Planner

Helpful Hints

HINT #1 MAKING A CHANFER ON THE WHEELS

Make a jig to round the edges of your wheels by cutting a bolt (removing the hexagon) and assembling it, as shown. Use this jig with your drill press.

Please note: Drill a 1/4" diameter hole in your wheels to install on this jig. When you finish sanding the wheels, re-drill the 1/4" diameter holes, this time using drill bit as specified in your plans.

Materials needed to make this jig

(1) 1/4" diameter bolt, 1 3/4" long

(2) 1/4" int.diameter flat washers

(1) 1/4" diameter nut

Illustration 1

View of parts necessary to make the sanding jig.

Illustration 2

Insert unthreaded end of bolt into drill press. Tighten chuck and slide on washer, wheel and 2nd washer. Secure everything using the nut.

Illustration 3

You are now ready to make the chanfer on the wheel. Start your drill press at low to medium r.p.m.s. Then, use a wood file to round the edges of your wheel. Use sandpaper to get a smooth finish.

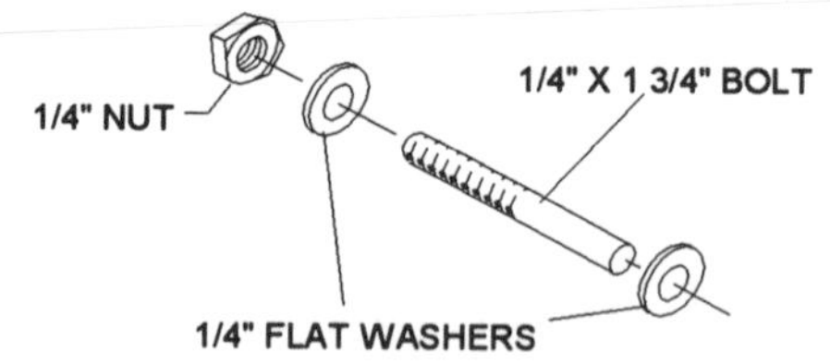

Illustration 1

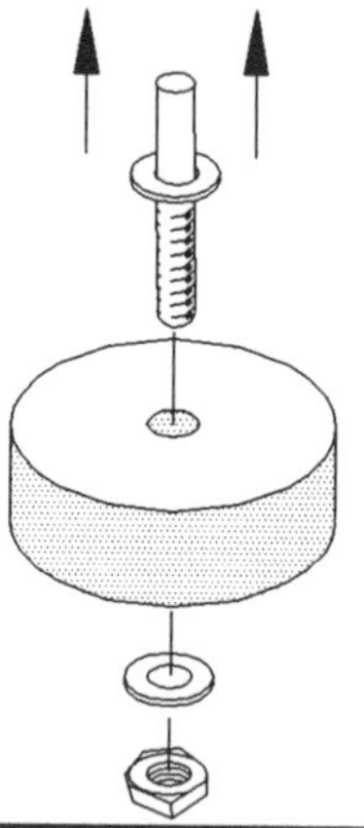

Illustration 2

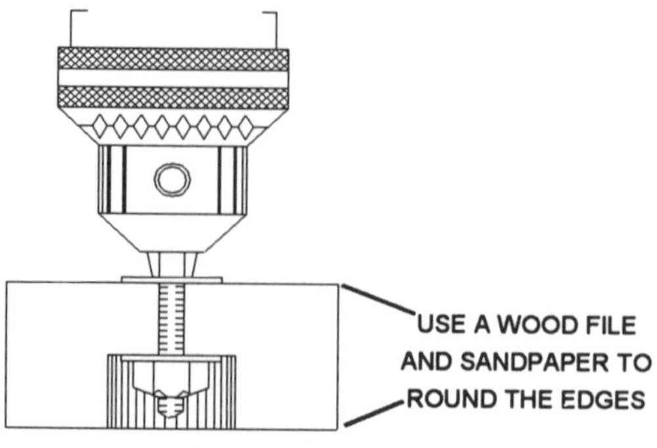

Illustration 3

HINT # 2 HOW TO MAKE A TRACK ASSEMBLY

Step 1

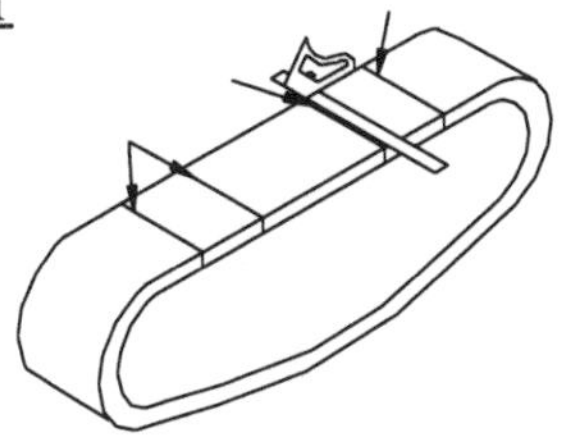

Trace four lines across surface, as shown.See hint #4.

Step 2

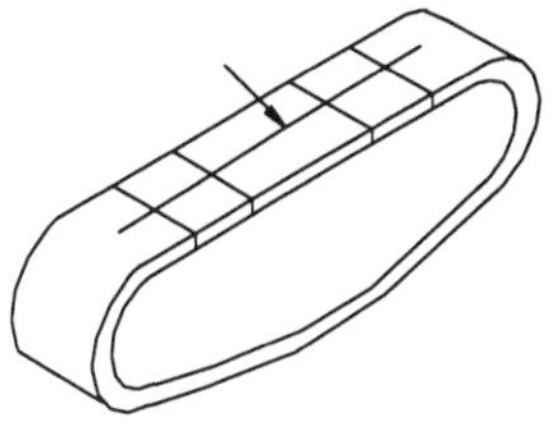

Trace a line down the centre, as shown.

Step 3

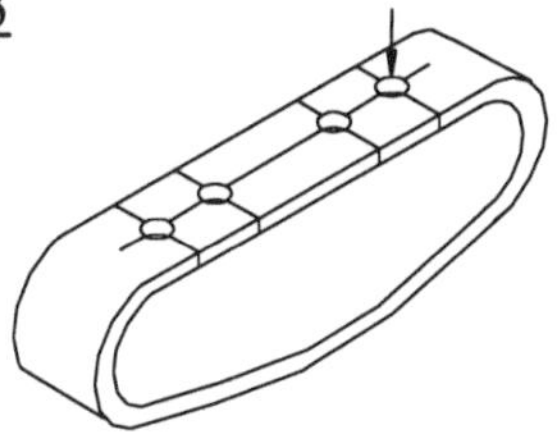

Drill four 1/2" holes on centre marks through first surface only.

Step 4

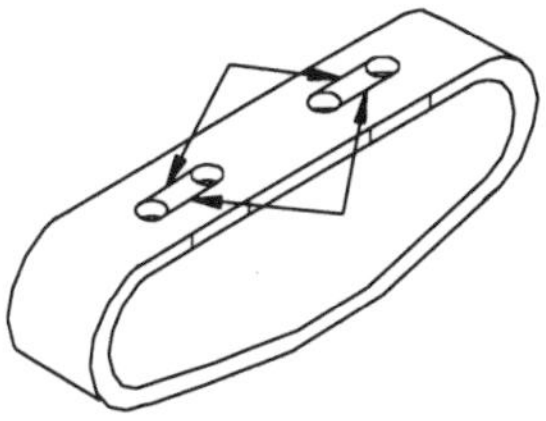

Using a sharp knife, cut lines tangent to holes, as shown.

Step 5

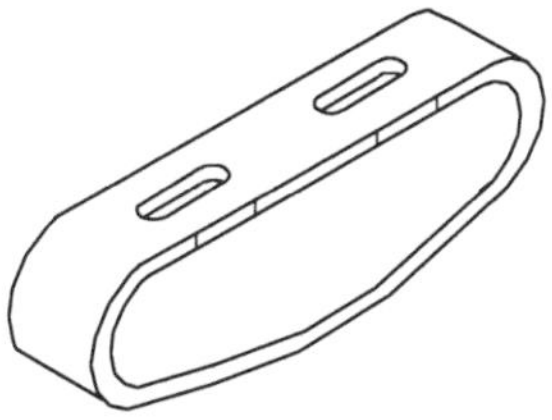

Opening properly cut.

HOW TO DRILL HOLES TO THE SPECIFIED DEPTH

To get holes to specified depth, use a piece of masking tape (1/2" wide or less). Wrap it around drill bit, as shown.

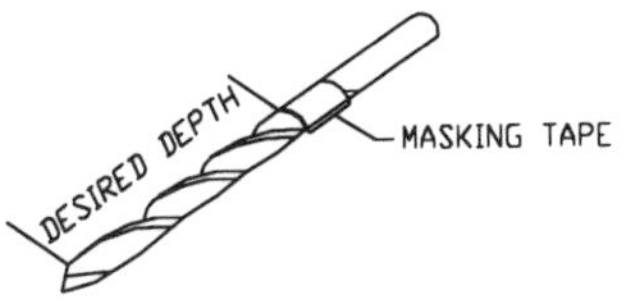

HINT # 3 MAKING A SHOVEL - THE EASY WAY

Step 1

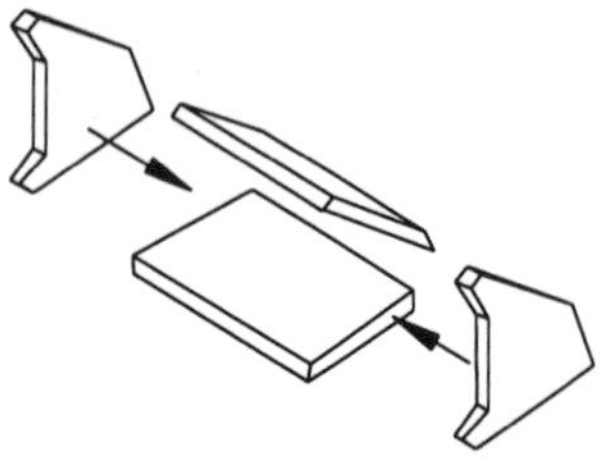

Glue shovel sides to shovel, top and bottom, as shown.

Step 2

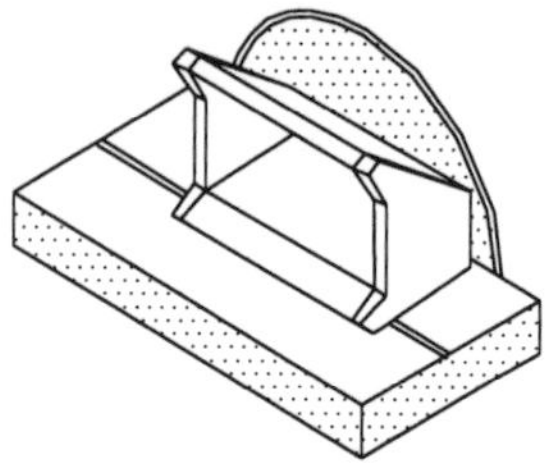

Sand the back surface to get a straight face.

Step 3

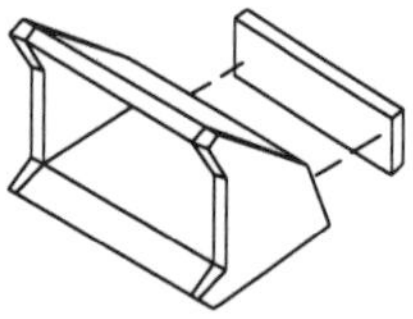

Glue on shovel back. Note: Shovel top, bottom and back are oversized to allow sanding.

Step 4

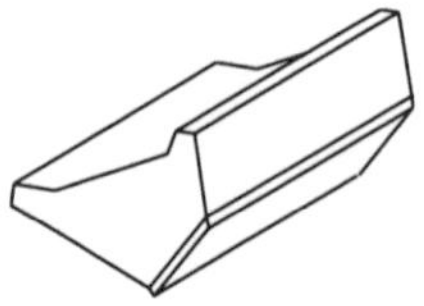

Do a final sanding to get your finished shovel.

HINT # 4 TRANSFERING THE GUIDELINES FROM PATTERNS

IMPORTANT: This section applies to these models only:

- Dozer
- Dozer Loader
- Excavator

Before removing pattern from material, transfer guidelines, as shown below.

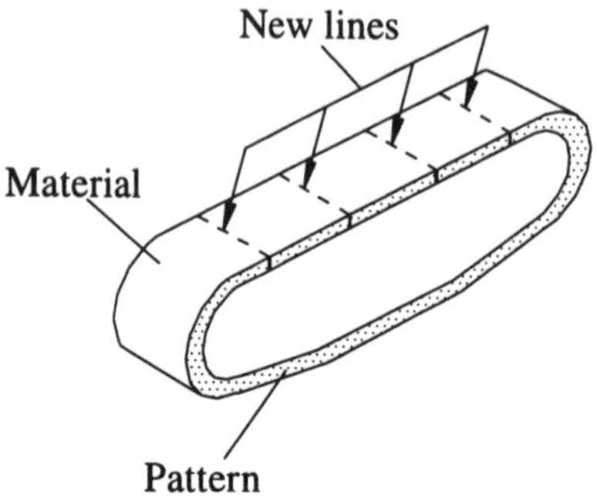

HINT #5 MAKE YOUR OWN SPECIAL DRILLING JIG

This homemade jig is used to hold the dowel in position while drilling a hole in the centre. This special jig is mainly used to make pins.

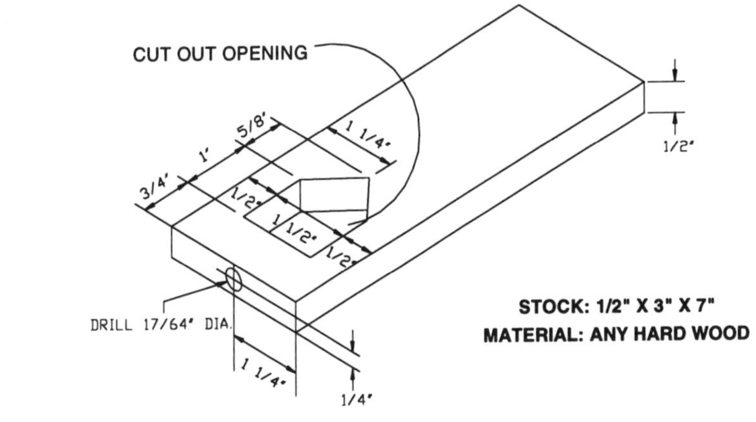

Illustration 1

Materials needed to make this jig

(1) 1/4" diameter rod, 3" long

(2) 1/4" wing nut

Illustration 1

Start by cutting a 1/2" thick stock to 3" wide x 7" long. Cut out opening and drill 17/64" diameter hole, as shown.

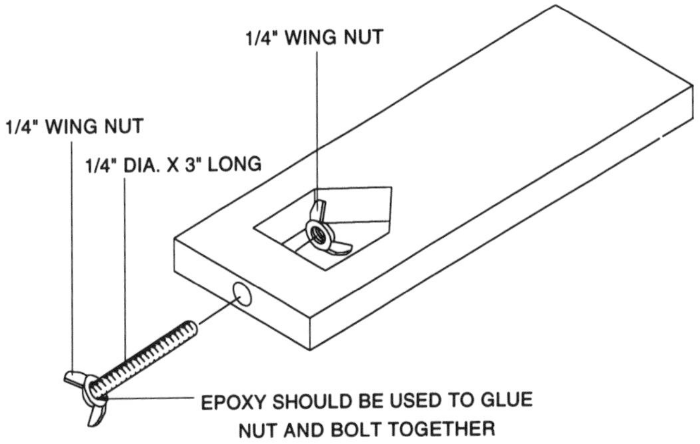

Illustration 2

Illustration 2

Assembly instructions

Illustration 3

Drawing showing how to use this jig.

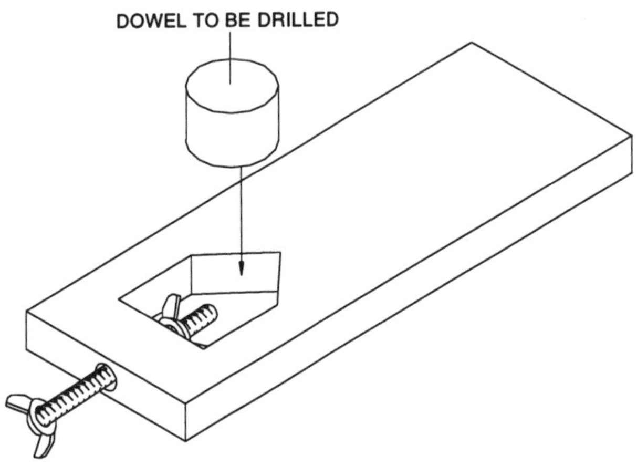

Illustration 3

IMPORTANT NOTE REGARDING THE FABRICATION OF PINS

The pins required to make the models shown in this book are made by using maple dowels.

Their diameters are not always consistent which means that you will need to sand them to get proper results.

LOADER

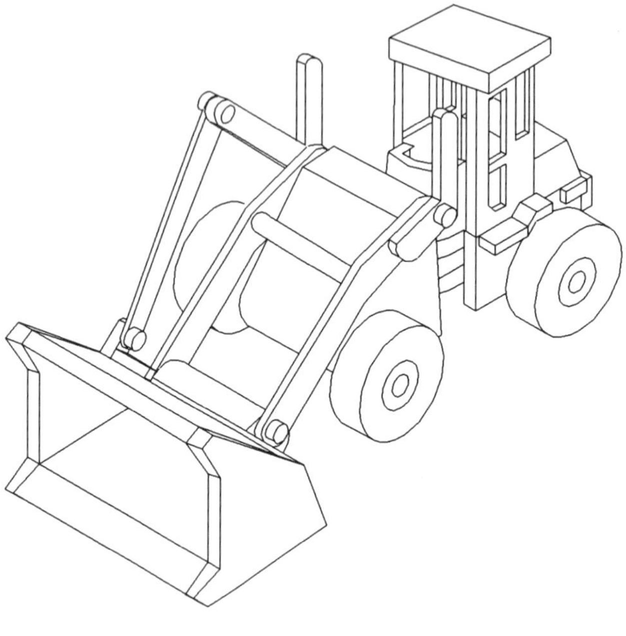

General Instructions - Loader

1- Start by cutting materials needed by following the list of materials, paying attention to the rough and finished size. **Identify the parts as they are cut.**

Please note: Different types of wood can be used for the various parts. It is suggested, however, that hard wood be used, since many of the parts would be much too fragile if using soft wood. We have used a combination of pine, maple and oak to give the models a nice contrast!

2- Remove the full-size patterns found in the appendix. Cut them out, leaving approximately 1/16" all around, and place on the proper piece of wood. Patterns can be secured to wood using either spray adhesive or rubber ciment. If using the latter, cut and sand the part first to finished size. If drilling is required, mark the hole by inserting a scriber or nail through the pattern into the wood. Remove the pattern before drilling.

You should have no trouble determining which surface to attach most of the patterns. Some parts, however, can be confusing since the pattern could fit on more than one surface. The drawings below indicate exactly which surface to attach the patterns for these parts.

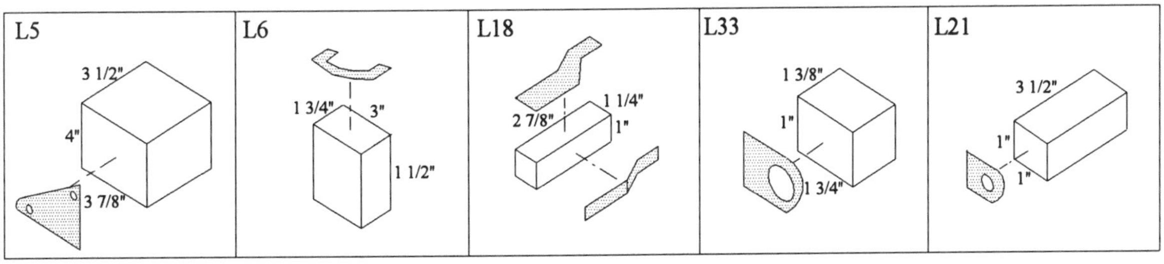

3- Look at the full-size drawing sheets to finish parts L10 and L23.

4- Parts L33 and L18 will need additional cuts and details, please refer to the additional information pages, to complete these parts.

5- Using maple dowels, make all pins, shafts, etc.

6- Follow the assembly drawings to complete your model.

List of Materials - Loader

Part	T	W	L	Material	Qty.	*
L1	3/8"	2 3/4"	4"	oak	1	R
L2	3/8"	2 3/4"	4"	oak	1	R
L3	3/8"	3"	4"	oak	1	R
L4	3/8"	3 1/4"	3 7/8"	maple	1	R
L5	4"	3 1/2"	3 7/8"	pine	1	F
L6	1 1/2"	3"	1 3/4"	pine	1	R
L7	3/8"	2 3/8"	6"	pine	2	R
L8	1/4"	1 7/8"	6 1/2"	pine	1	R
L9	3/4"	3"	7 5/8"	pine	2	R
L10	3/4"	2 13/16"	6 1/8"	pine	1	F
L11	3/4"	1 1/2"	2 3/4"	maple	1	R
L12	3/8"	2 5/8"	5 5/8"	pine	1	F
L13	3/8"	3 1/2"	5 5/8"	pine	1	F

Part	T	W	L	Material	Qty.	*
L14	3/8"	3 3/8"	3 5/8"	pine	2	R
L15	1/4"	5/8"	1 1/4"	oak	2	F
L16	3/8"	2 3/4"	3 5/8"	pine	1	F
L17	1/2"	1 3/4"	4 1/4"	oak	1	R
L18	1"	1 1/4"	2 7/8"	pine	2	R
L19	1 1/2"	3 9/16"	3 3/8"	pine	1	F
L20	1/4"	1 1/8"	6 3/4"	maple	1	R
L21	1"	1"	3 1/2"	maple	1	F
L22	3/8"	2 1/8"	7 3/8"	maple	2	R
L23	1 1/4"	3 1/8" DIA.		oak	4	F
L24	3/8"	3 1/2"	4 5/8"	maple	1	R
L33	1"	1 3/8"	1 3/4"	maple	1	F

R = Rough size
F = Finished size

T = Thickness
W = Width
L = Length

Instructions:

R= Rough sizes, the material is cut oversized so you have ample room to apply the pattern on the surface. Sanding is not required at this point.

F = Finished Size: Cut and sand parts to finished size.

Full-Sized Patterns: Set One

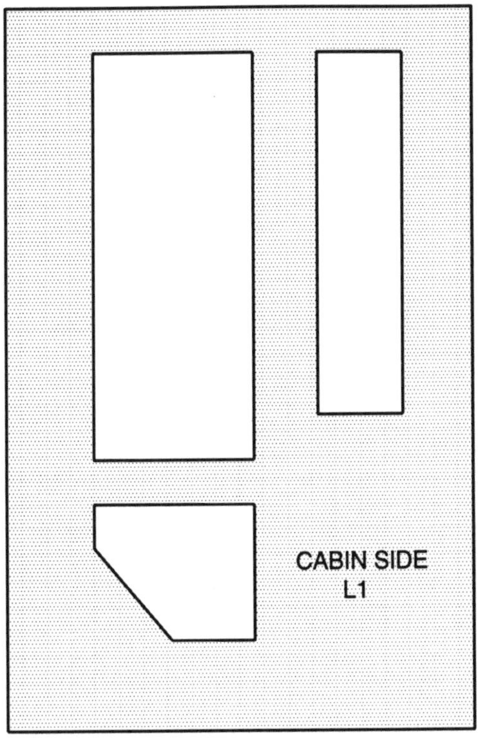

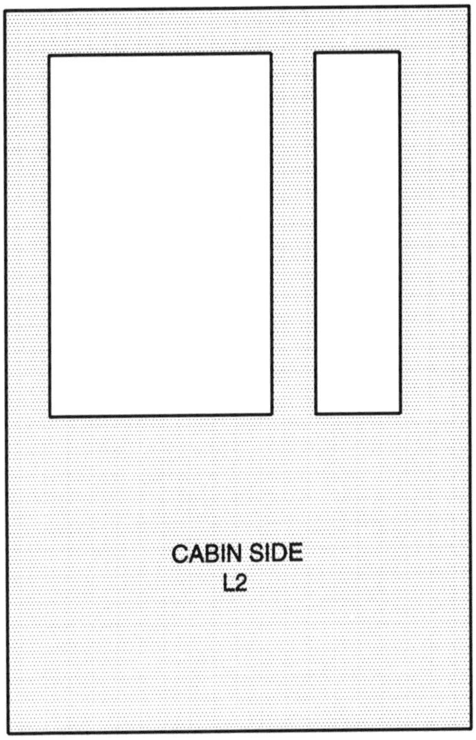

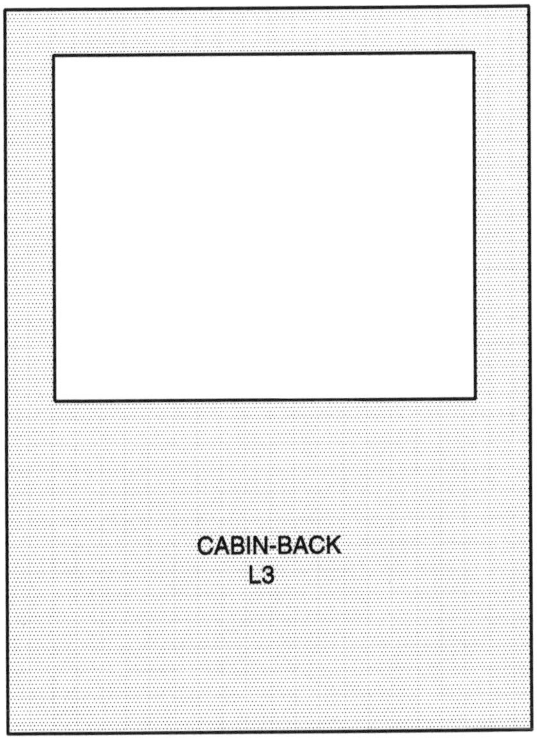

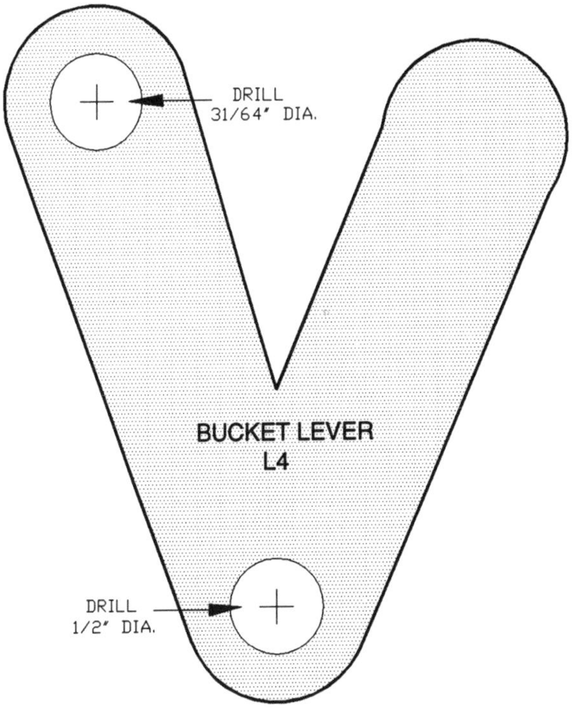

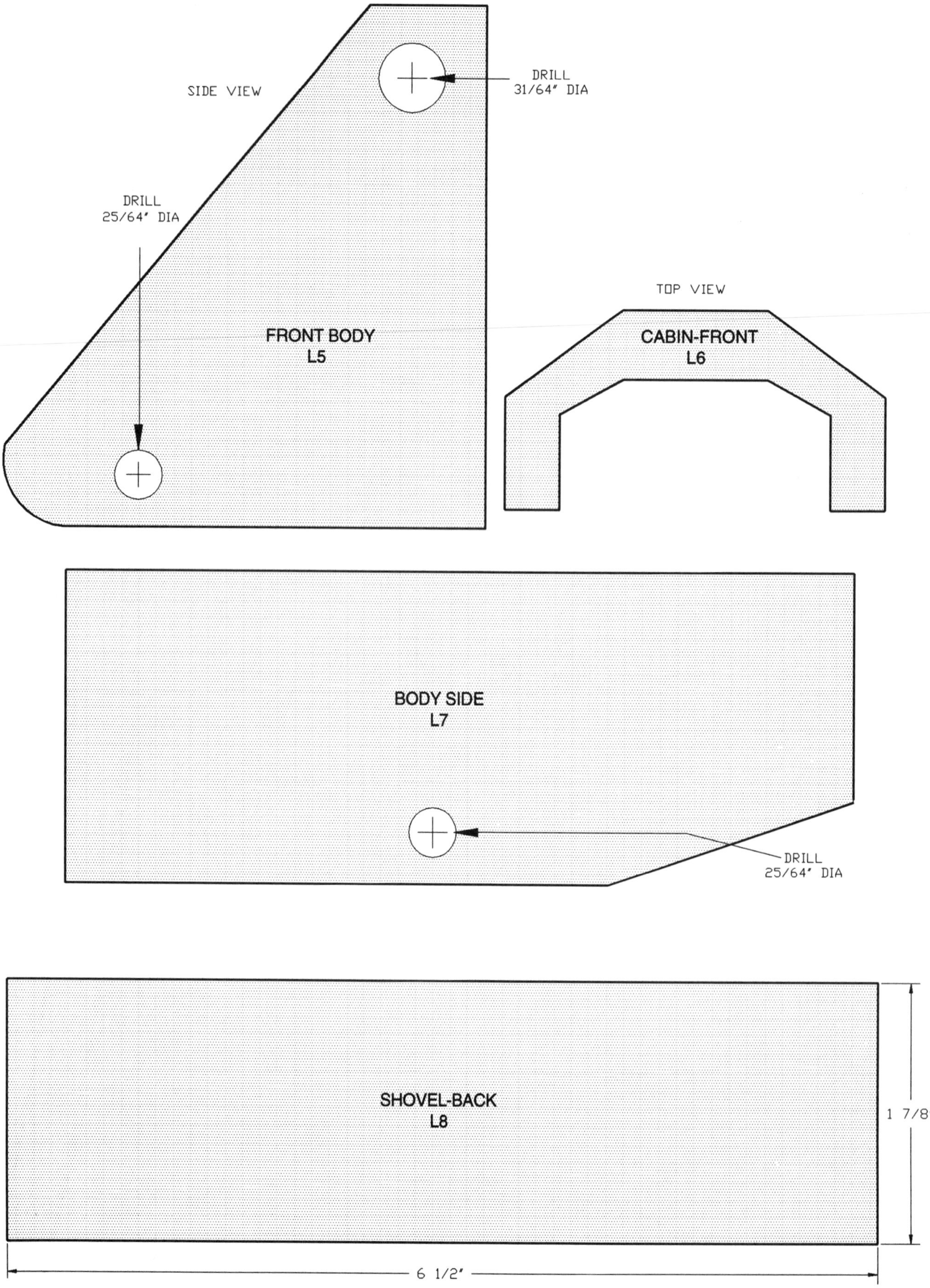
SIDE VIEW
DRILL
31/64" DIA
DRILL
25/64" DIA
FRONT BODY
L5
TOP VIEW
CABIN-FRONT
L6
BODY SIDE
L7
DRILL
25/64" DIA
SHOVEL-BACK
L8
1 7/8"
6 1/2"

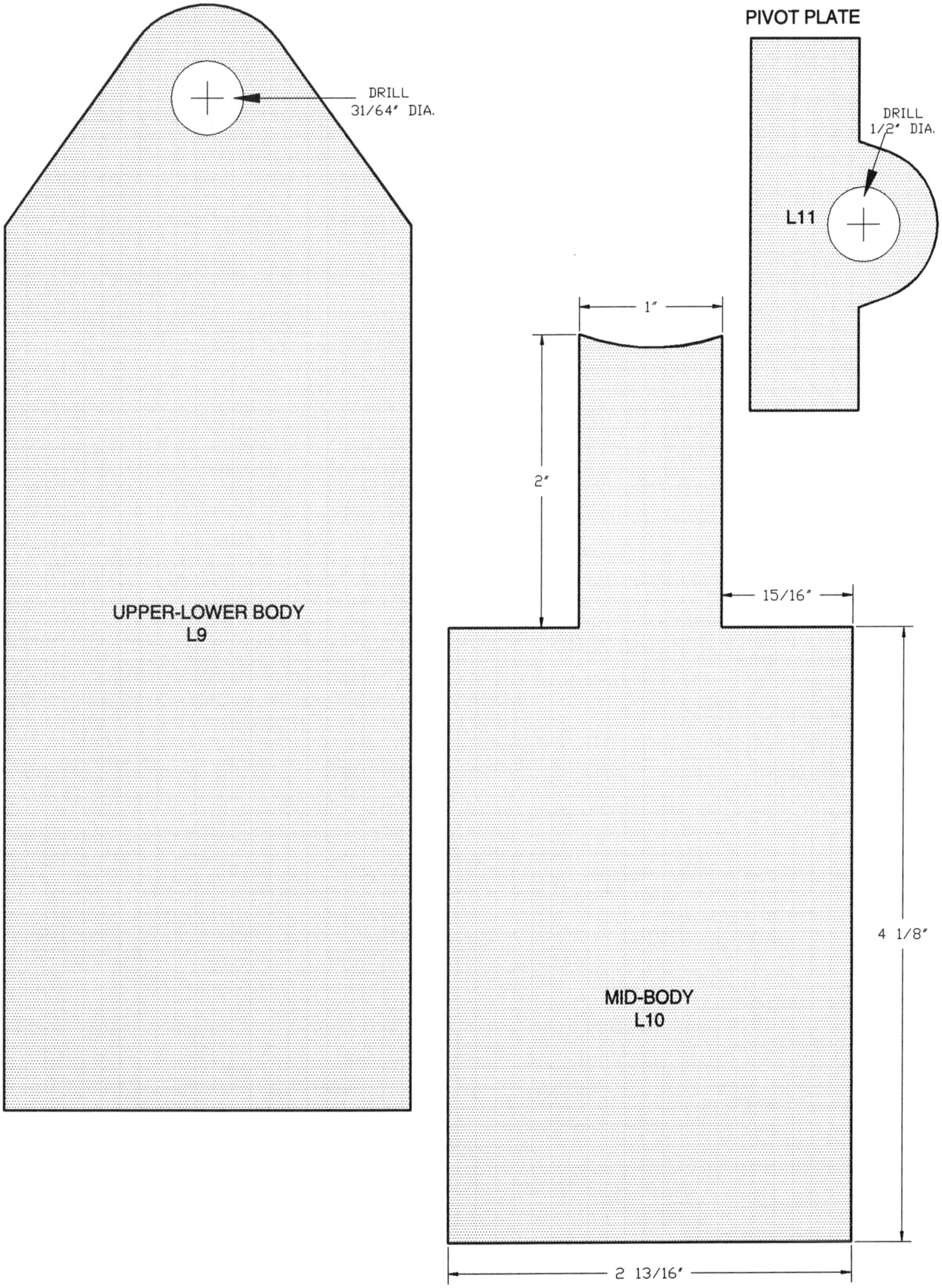
DRILL
31/64" DIA.
UPPER-LOWER BODY
L9
PIVOT PLATE
DRILL
1/2" DIA.
L11
1"
2"
15/16"
4 1/8"
MID-BODY
L10
2 13/16"

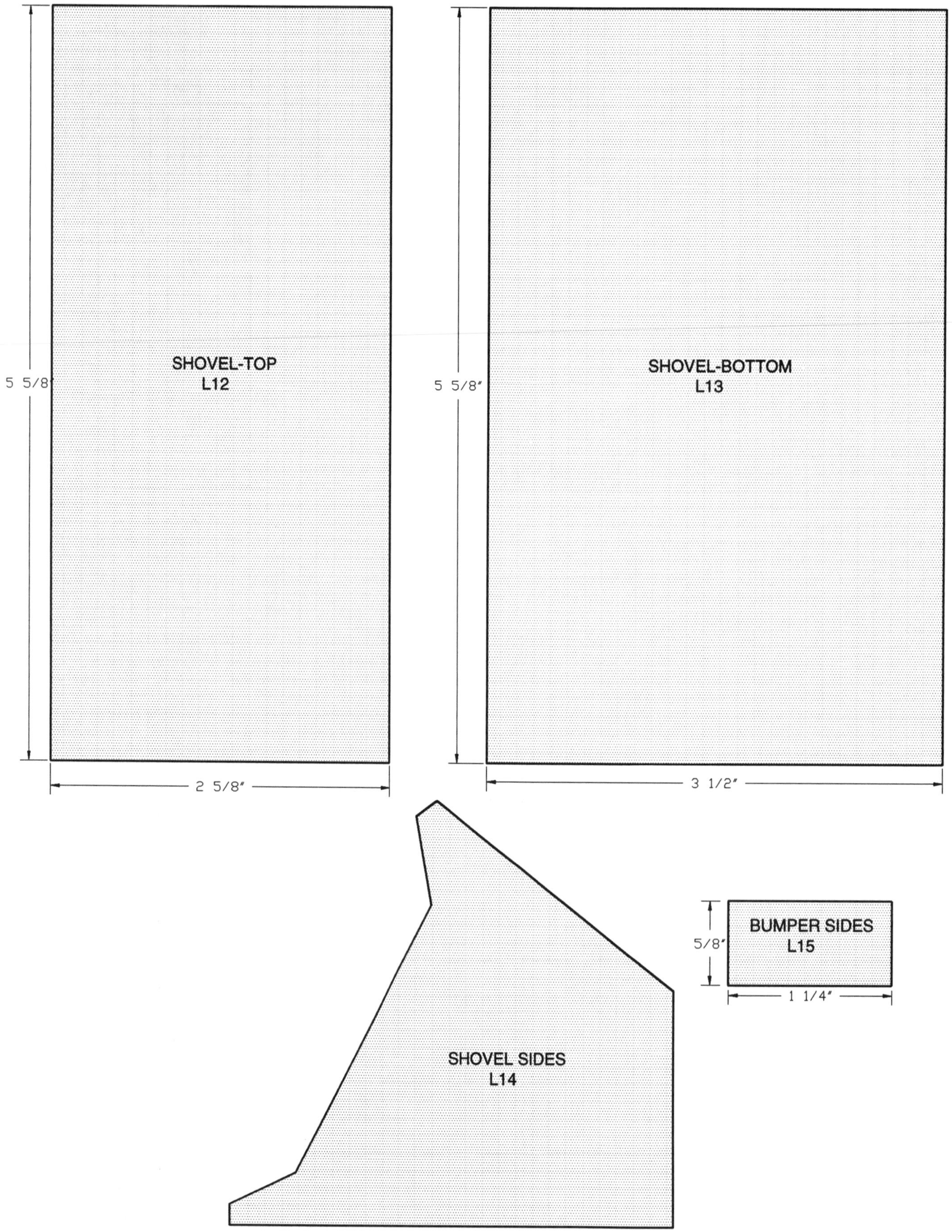
5 5/8
SHOVEL-TOP
L12
2 5/8″
5 5/8″
SHOVEL-BOTTOM
L13
3 1/2″
SHOVEL SIDES
L14
5/8″
BUMPER SIDES
L15
1 1/4″

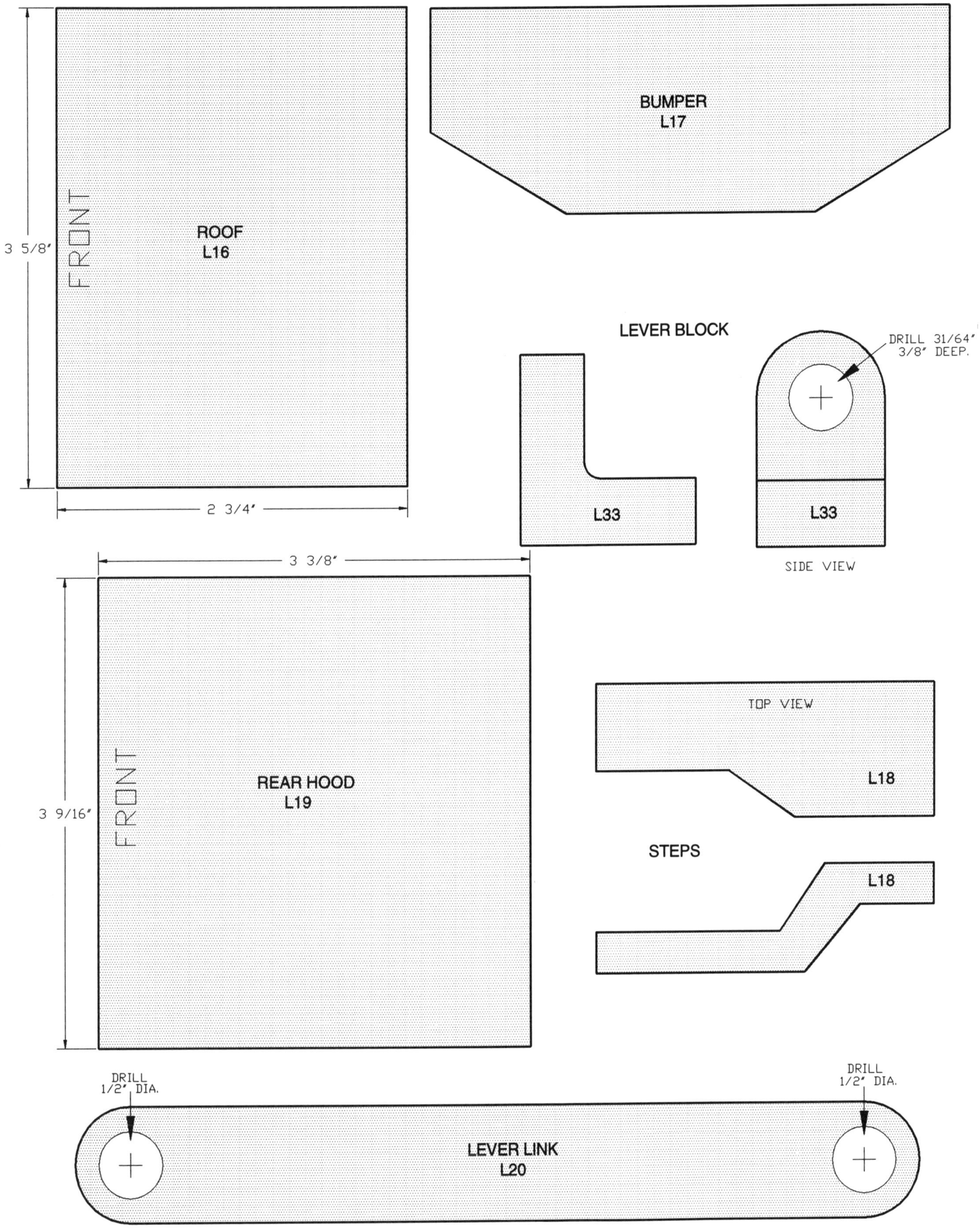
BUMPER
L17
3 5/8"
FRONT
ROOF
L16
2 3/4"
LEVER BLOCK
DRILL 31/64"
3/8" DEEP.
L33
L33
SIDE VIEW
3 3/8"
TOP VIEW
L18
3 9/16"
FRONT
REAR HOOD
L19
STEPS
L18
DRILL
1/2" DIA.
DRILL
1/2" DIA.
LEVER LINK
L20

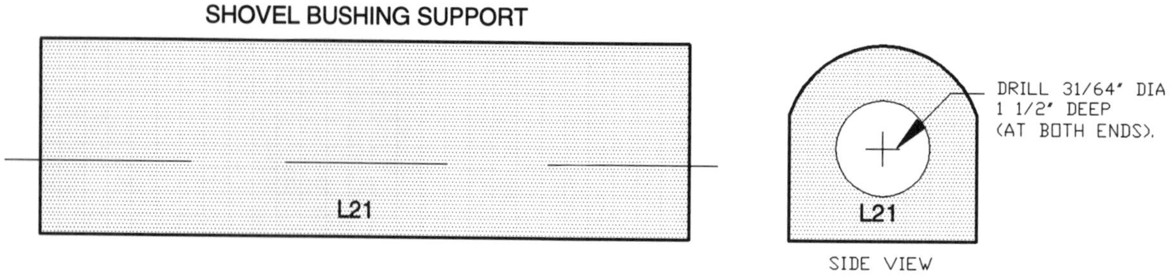
SHOVEL BUSHING SUPPORT
L21
DRILL 31/64" DIA
1 1/2" DEEP
(AT BOTH ENDS).
L21
SIDE VIEW

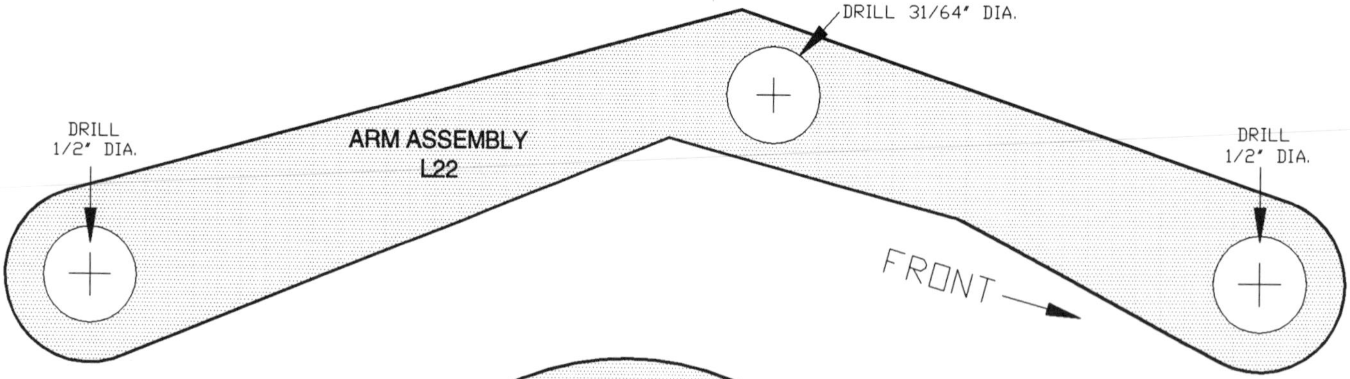
DRILL 31/64" DIA.
DRILL
1/2" DIA.
ARM ASSEMBLY
L22
DRILL
1/2" DIA.
FRONT

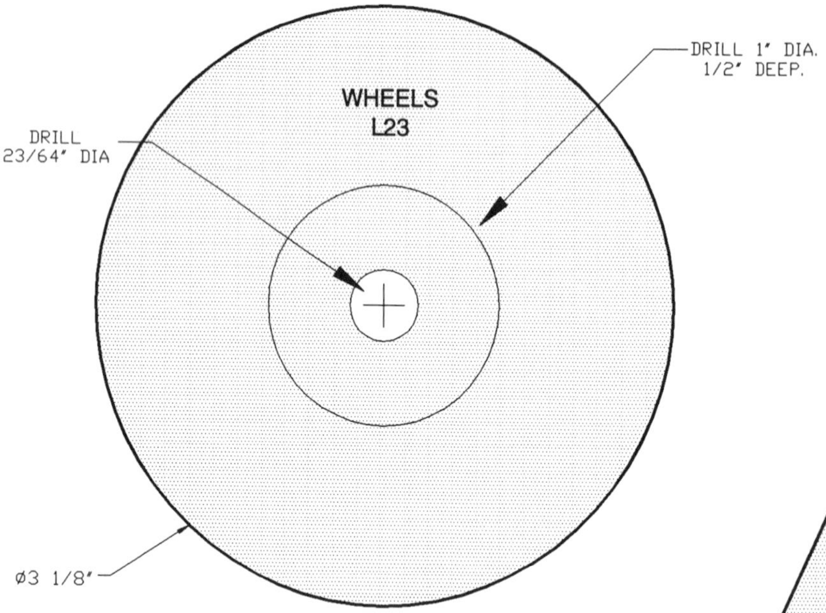
DRILL 1" DIA.
1/2" DEEP.
WHEELS
L23
DRILL
23/64" DIA
Ø3 1/8"

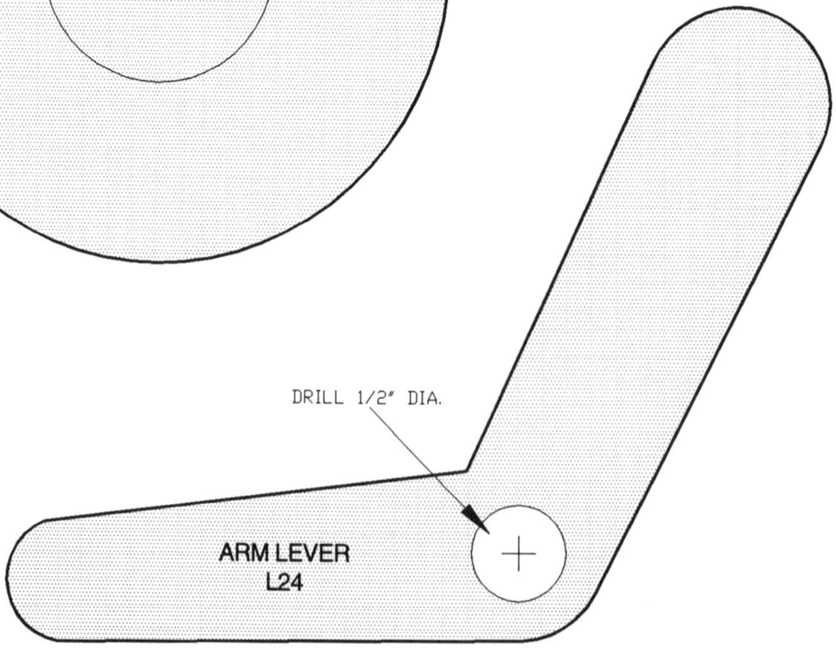
DRILL 1/2" DIA.
ARM LEVER
L24

L25

WHEEL SHAFT
(4 REQUIRED)

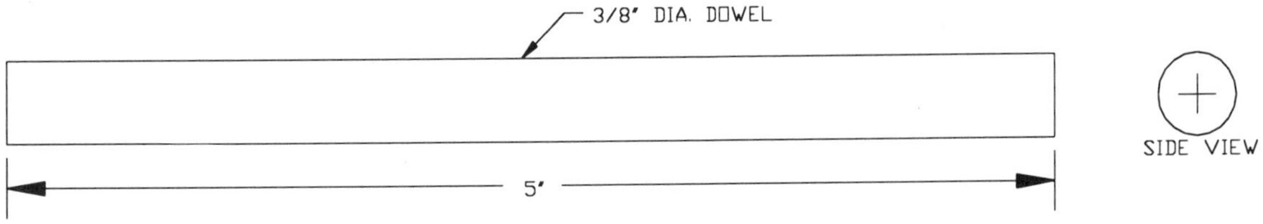

L28

RE-ENFORCING BAR
(1 REQUIRED)

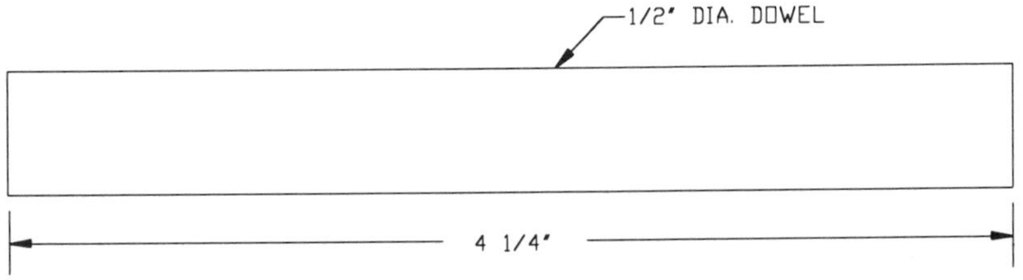

L29

MAIN LEVER PINS
(2 REQUIRED)

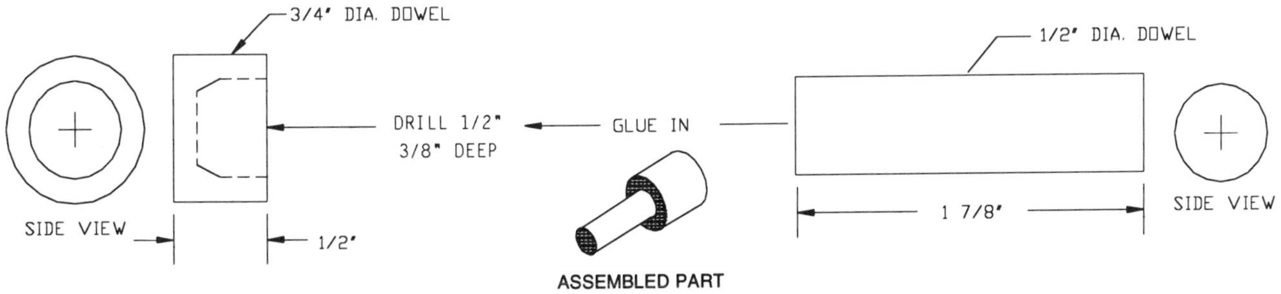

L30

MAIN LEVER PINS
(2 REQUIRED)

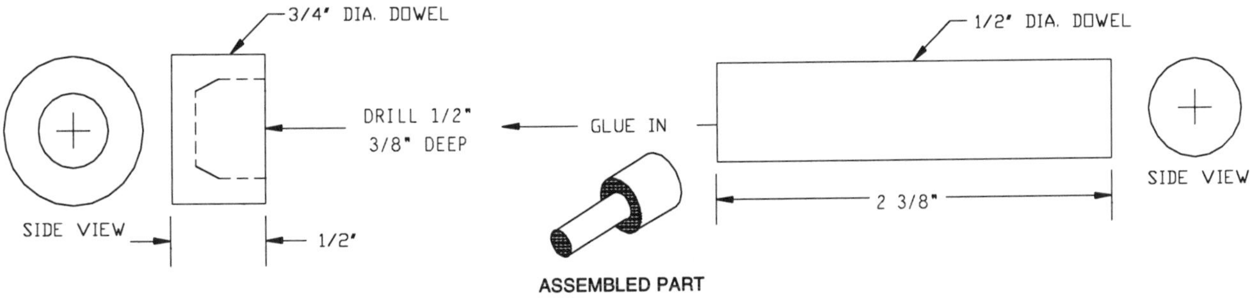

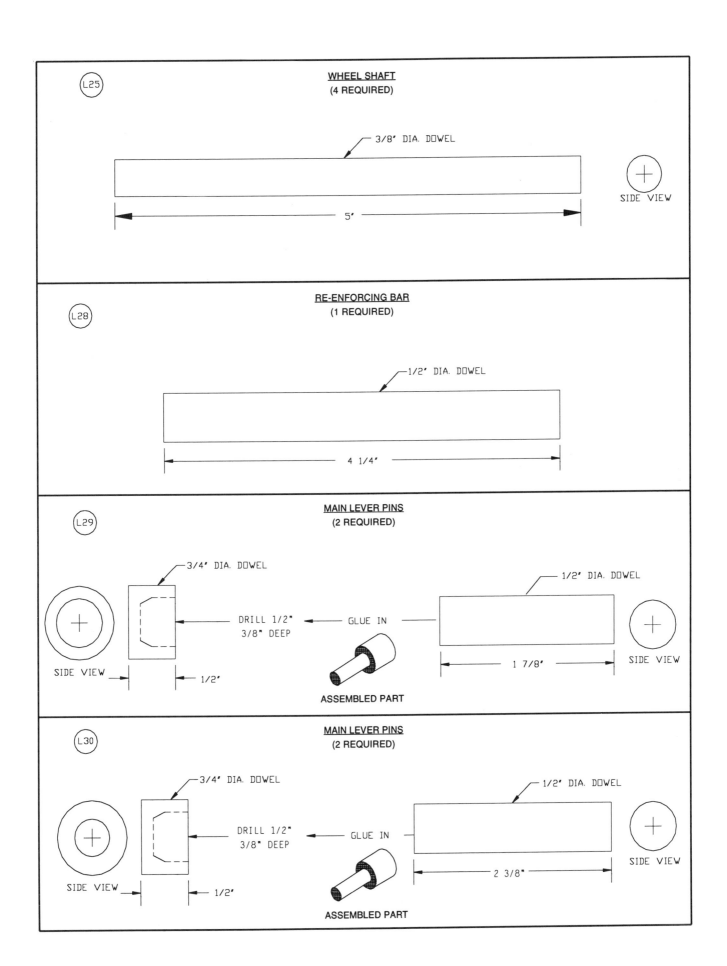

L31

LOCKING PINS

(2 REQUIRED)

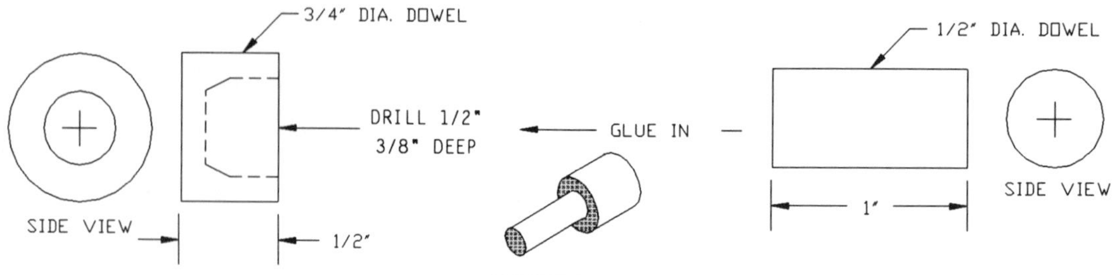

L32

MAIN PIN

(1 REQUIRED)

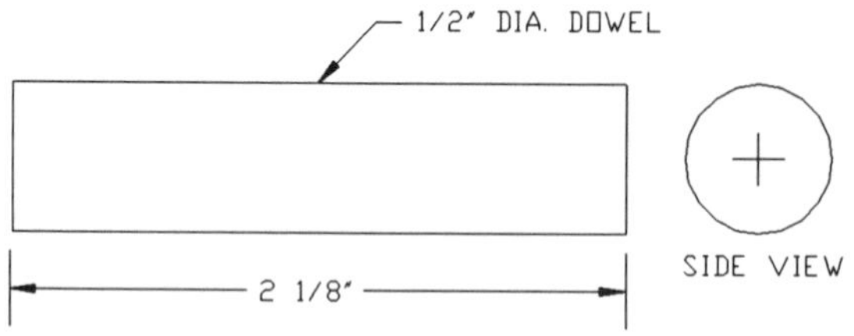

Additional Information - Loader

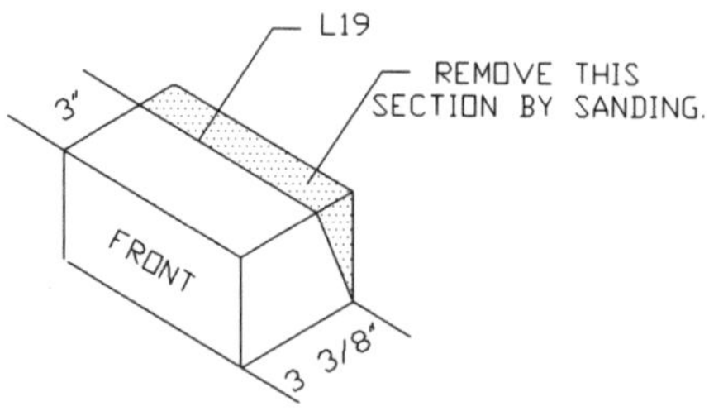

Sand angle on rear hood L19, as shown.

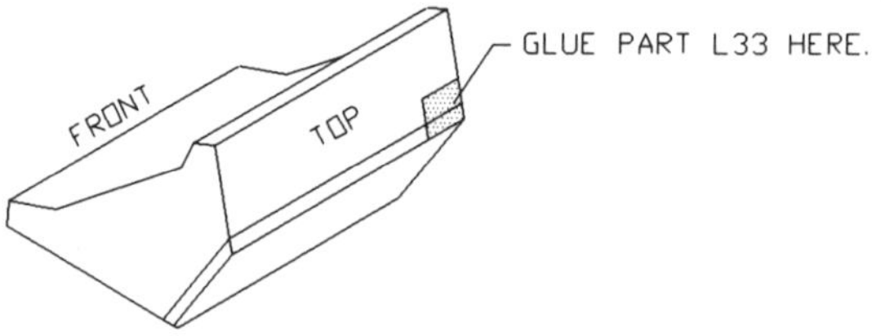

Where to glue part L33 onto shovel assembly.

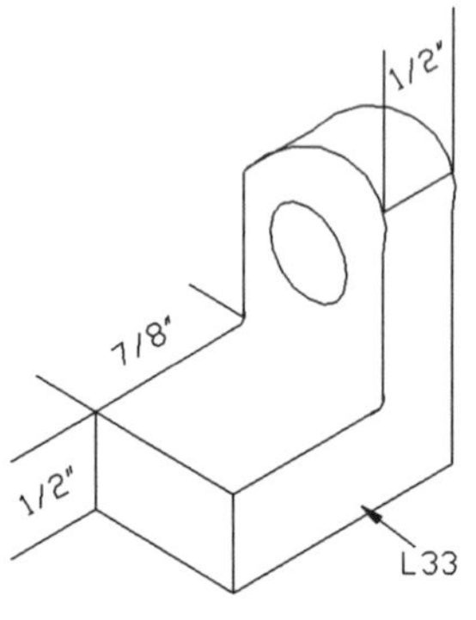

Finished part L33.

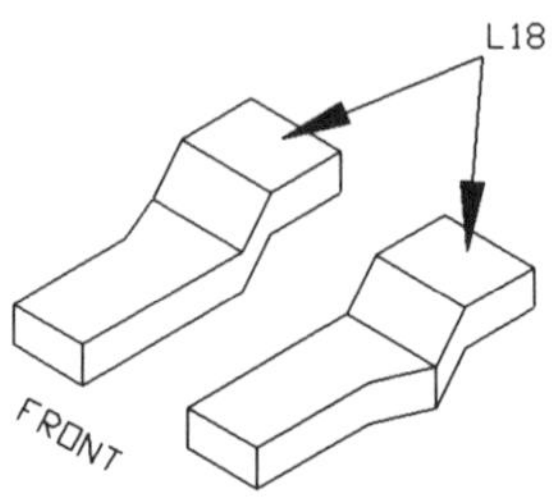

Finished parts L18.

Loader - Assembly Drawings

1

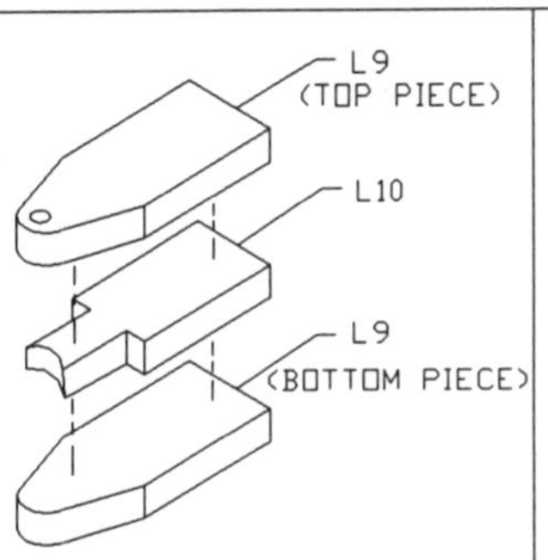

Glue mid-body L10 between upper and lower body L9.

2

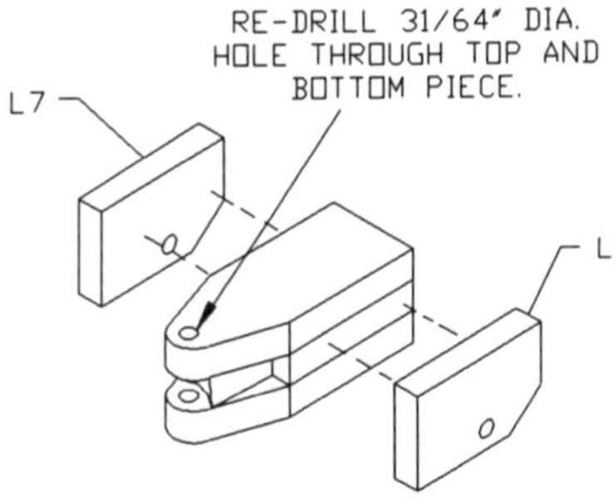

Re-drill hole, as shown. Glue sides L7 to body assembly.

3

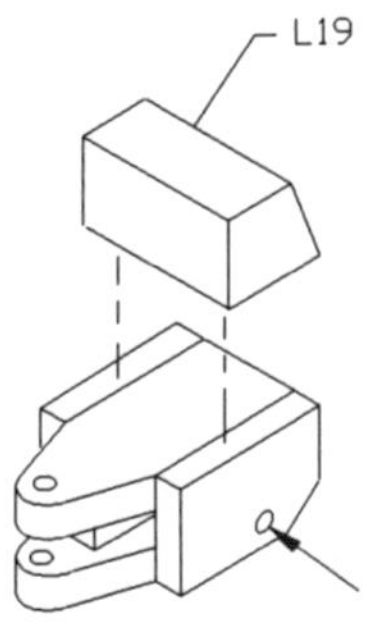

Re-drill 25/64" Dia. hole (from side to side). Glue rear hood L19.

4

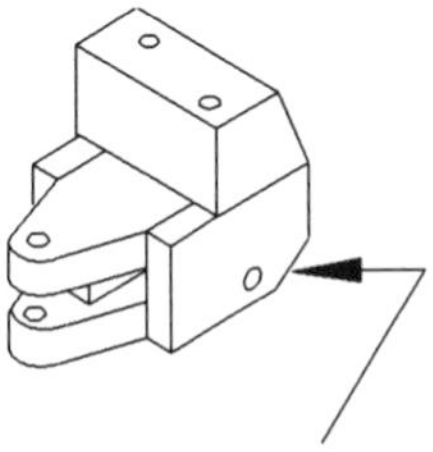

Re-sand angle (from side to side).

5

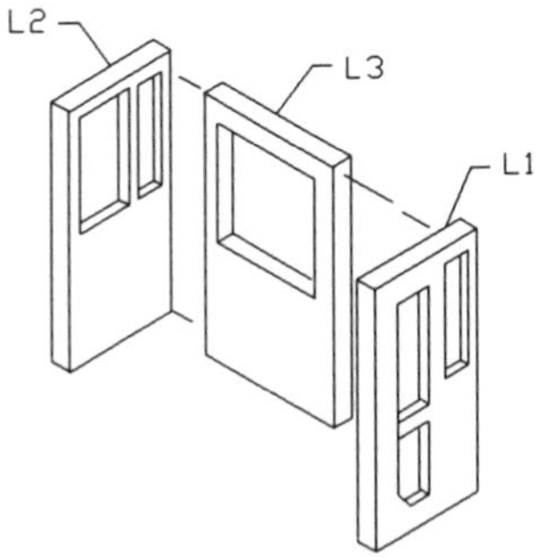

Glue cabin sides L1 & L2 to cabin back L3.

6

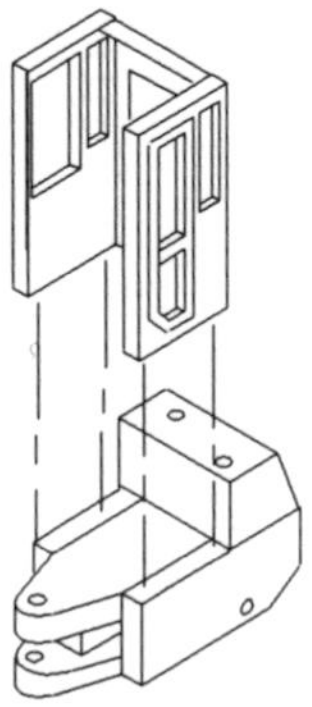

Glue cabin onto body, as shown.

7

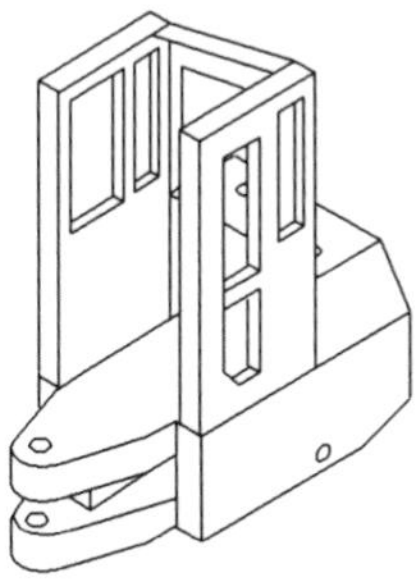

Cabin properly assembled.

8

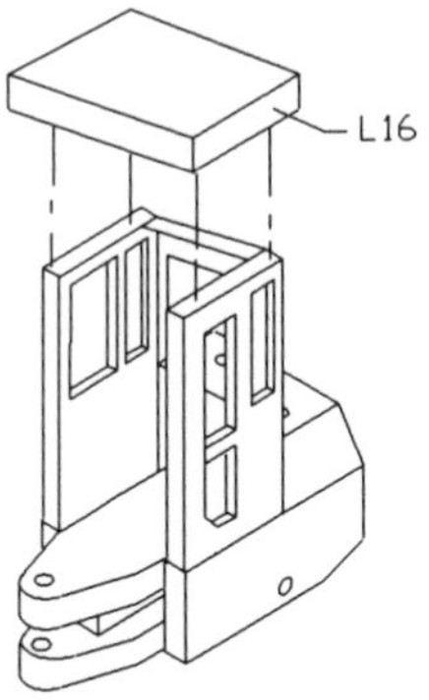

Glue roof L16 onto cabin.

9

IMPORTANT

Cabin must be completely assembled before glue dries so that you may make corrections for proper fit, if necessary.

10

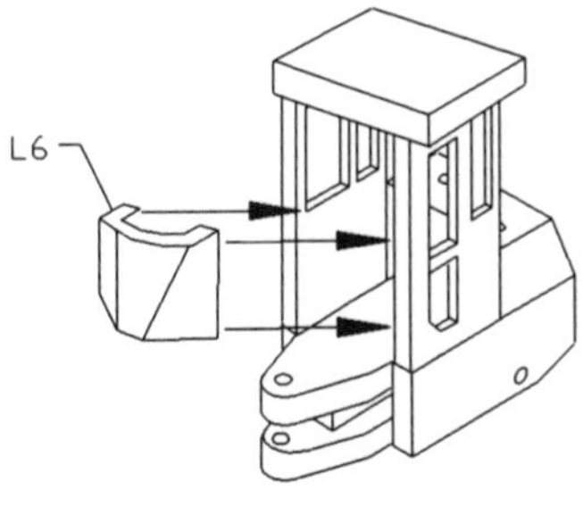

Glue cabin front L6, as shown.

11

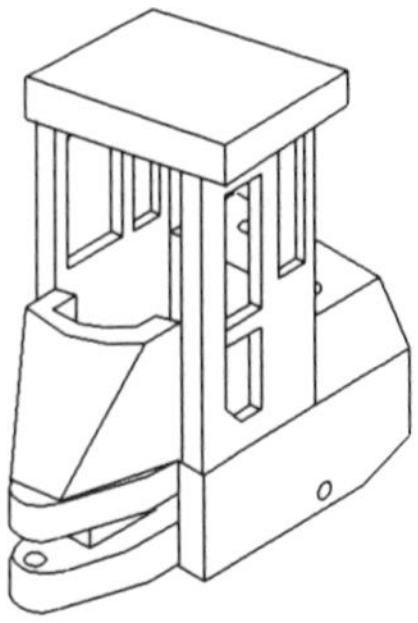

Start front body assembly.

12

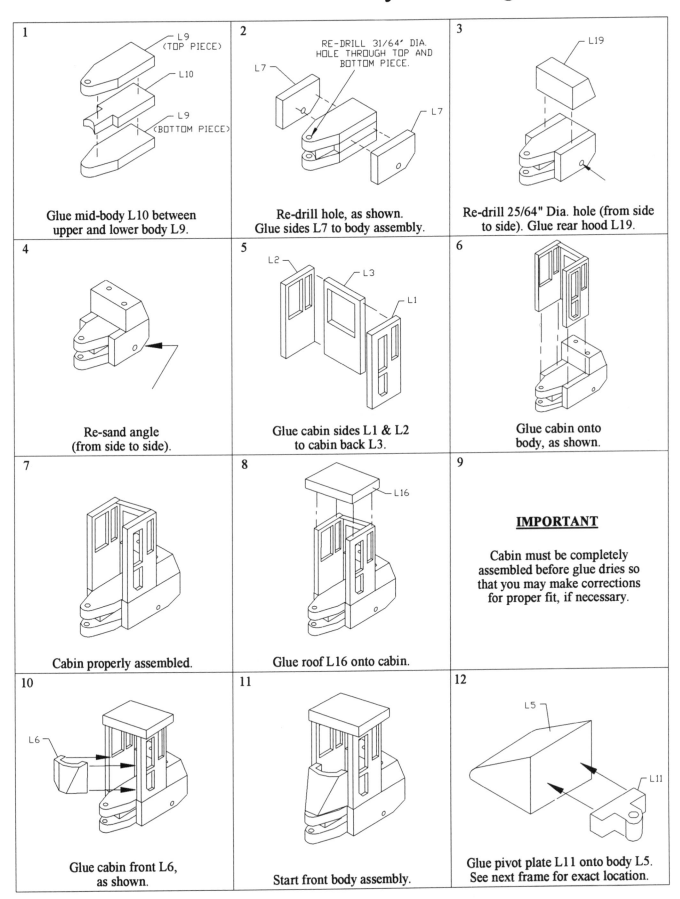

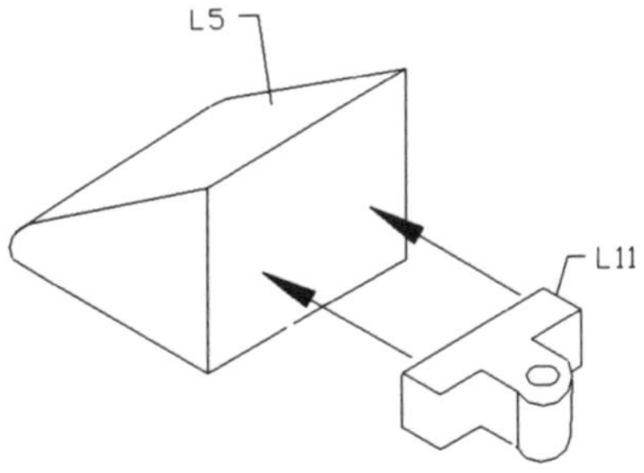

Glue pivot plate L11 onto body L5. See next frame for exact location.

13

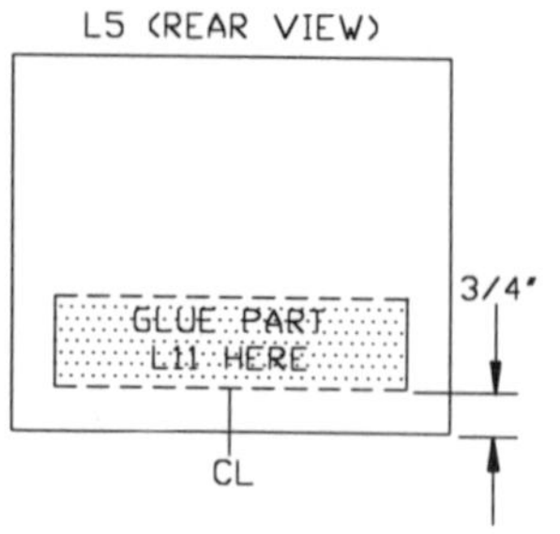

Where to glue part L11 onto L5.

14

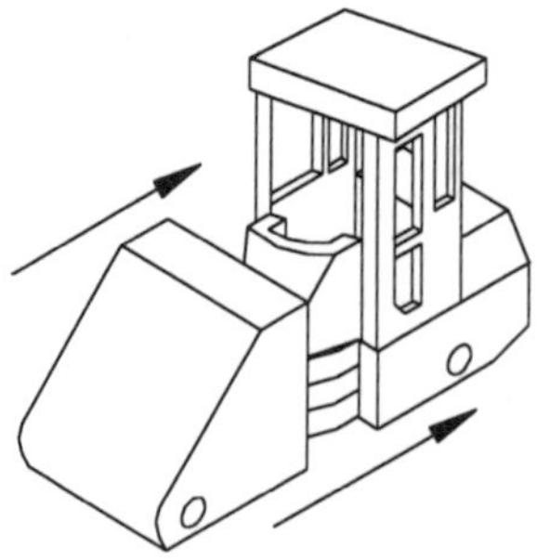

Slide front assembly into rear body.

15

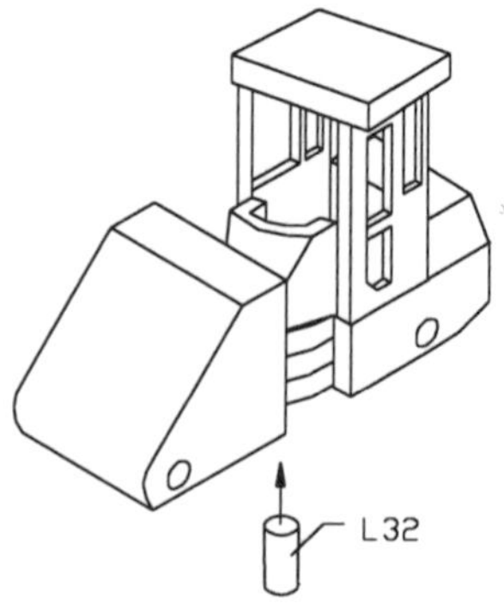

Insert main pivot pin L32.

16

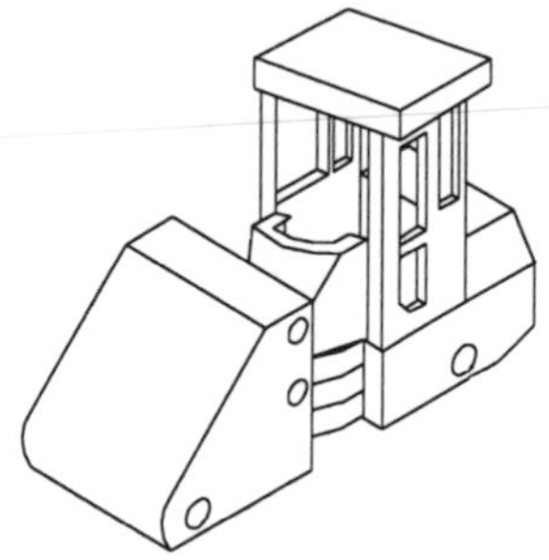

Model should pivot freely.

17

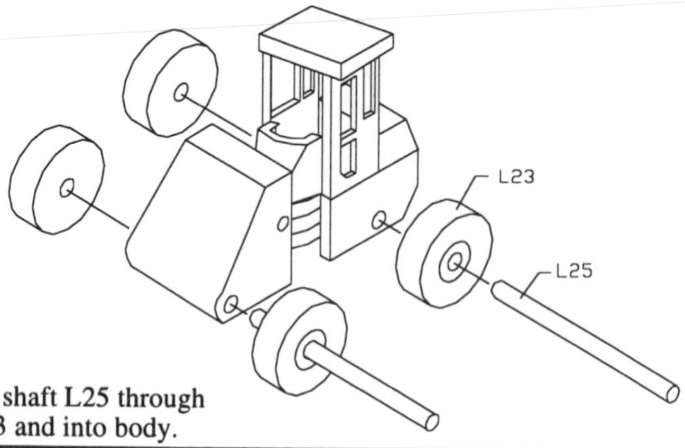

Insert wheel shaft L25 through wheels L23 and into body.

18

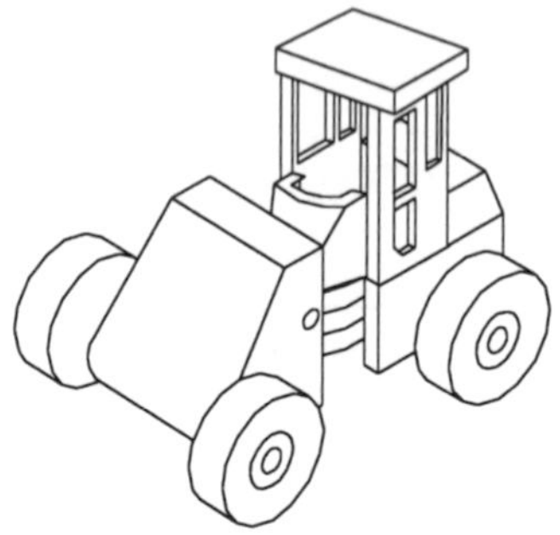

Start shovel assembly.

19

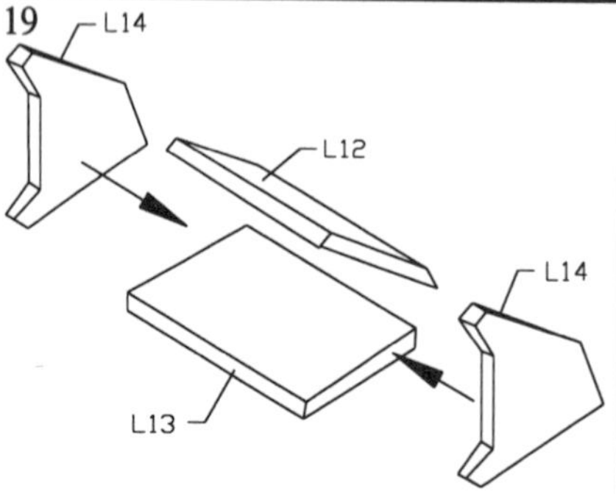

Glue shovel sides L14 to shovel top L12 and shovel bottom L13.

20

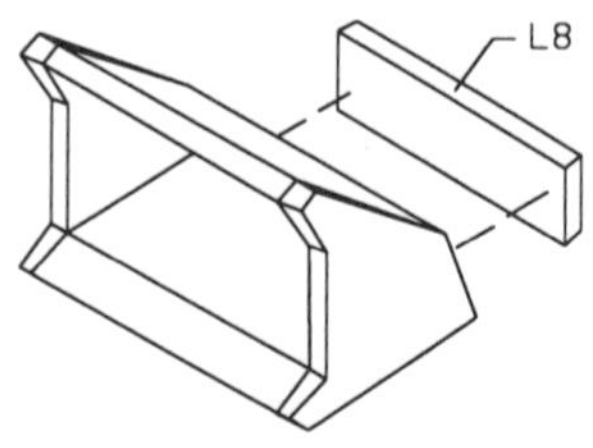

Glue shovel back L8. See page 6 for hints on making shovel.

21

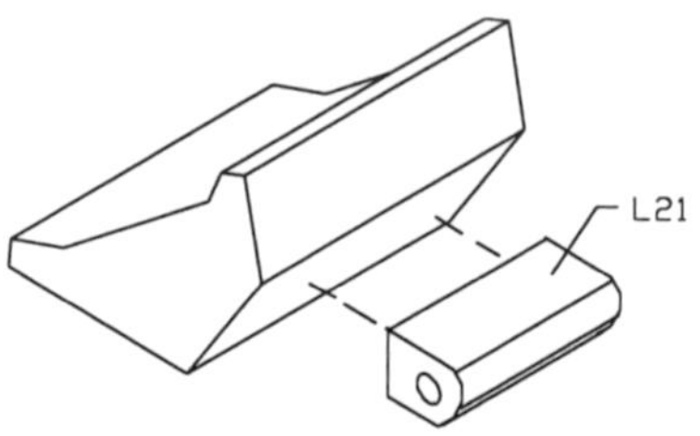

Glue part L21 centered onto shovel back.

22

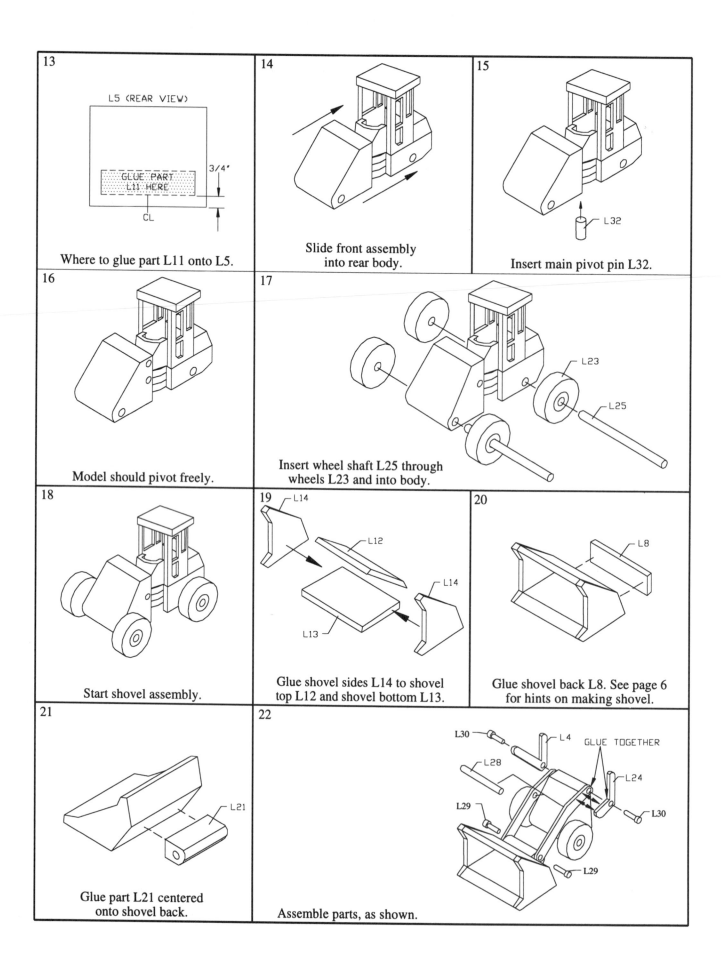

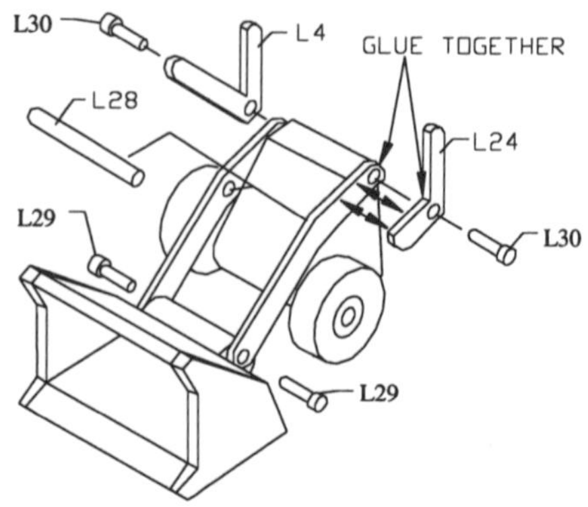

Assemble parts, as shown.

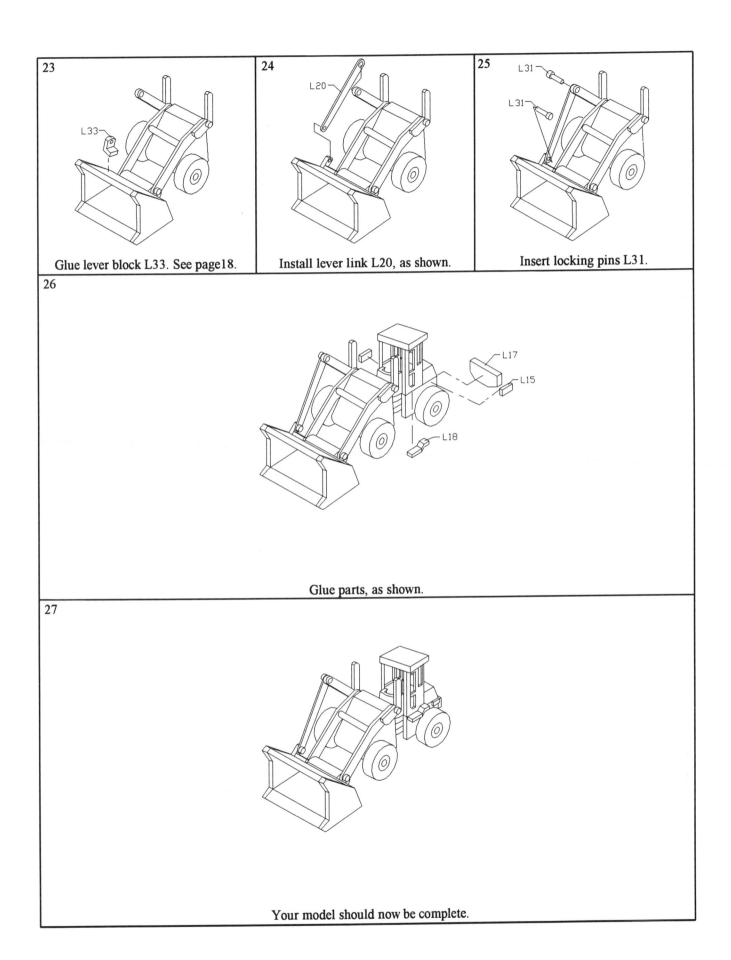

23

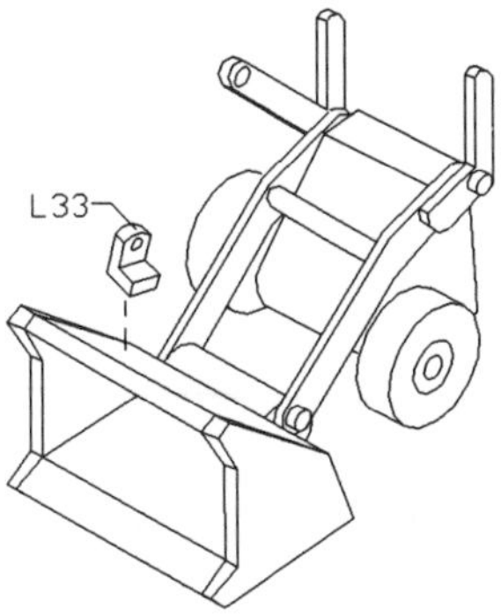

Glue lever block L33. See page18.

24

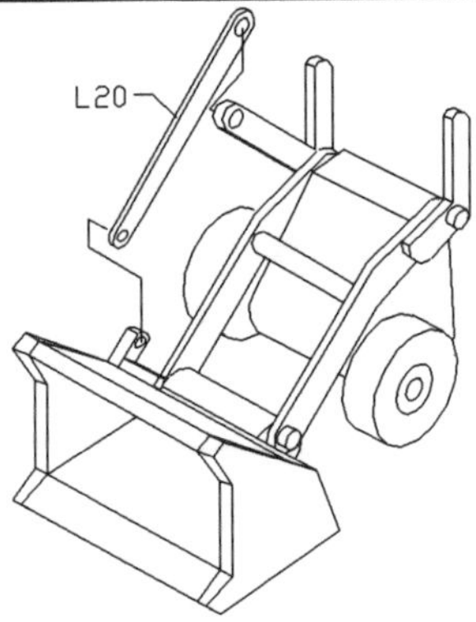

Install lever link L20, as shown.

25

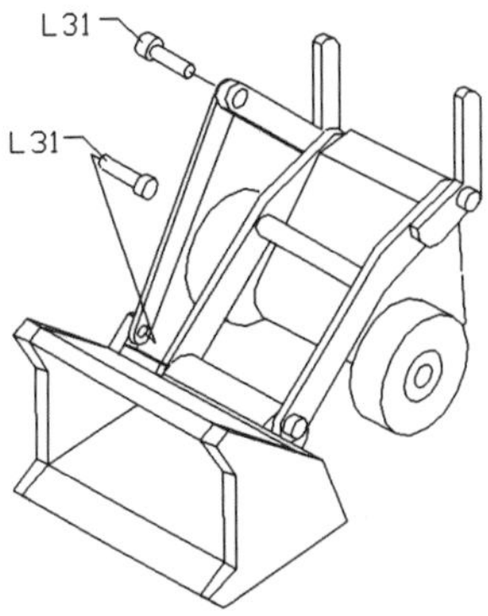

Insert locking pins L31.

26

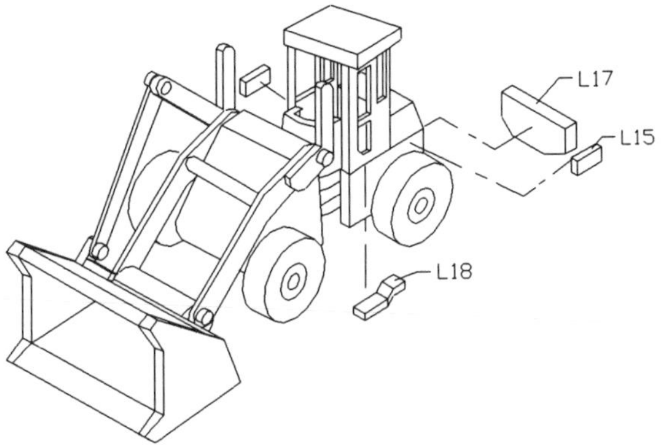

Glue parts, as shown.

27

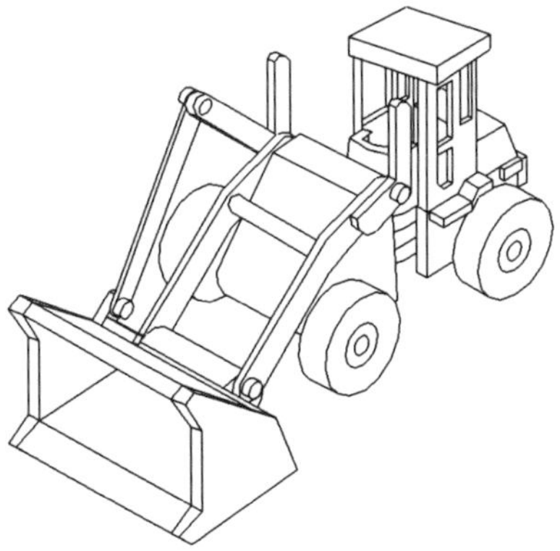

Your model should now be complete.

DOZER

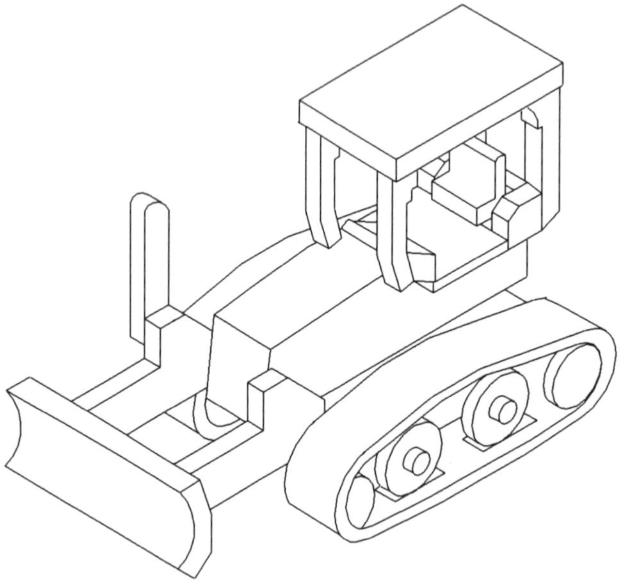

General Instructions - Dozer

1- Start by cutting materials needed by following the list of materials, paying attention to the rough and finished size. **Identify the parts as they are cut.**

Please note: Different types of wood can be used for the various parts. It is suggested, however, that hard wood be used, since many of the parts would be much too fragile if using soft wood. We have used a combination of pine, maple and oak to give the models a nice contrast!

2- Remove the full-size patterns found in the appendix. Cut them out, leaving approximately 1/16" all around, and place on the proper piece of wood. Patterns can be secured to wood using either spray adhesive or rubber ciment. If using the latter, cut and sand the part first to finished size. If drilling is required, mark the hole by inserting a scriber or nail through the pattern into the wood. Remove the pattern before drilling.

You should have no trouble determining which surface to attach most of the patterns. Some parts, however, can be confusing since the pattern could fit on more than one surface. The drawings below indicate exactly which surface to attach the patterns for these parts.

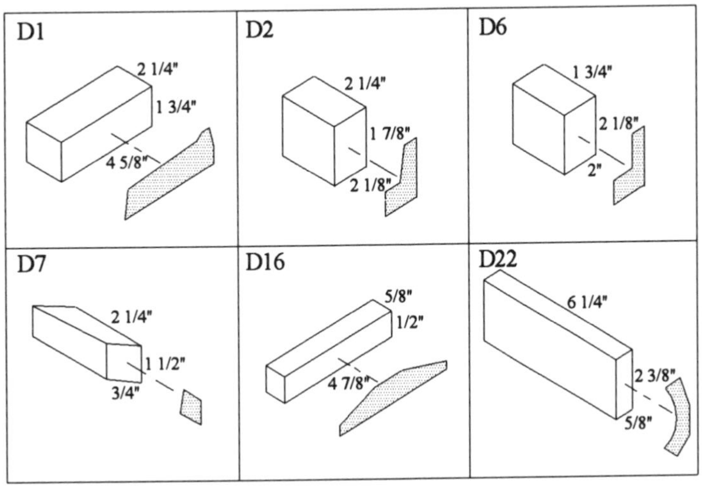

3- Look at the full-size drawing sheets to finish parts D3 and D21.

4- Using maple dowels, make all pins, shafts, etc.

5- Follow the assembly drawings to complete your model.

List of Materials - Dozer

Part	T	W	L	Material	Qty.	*
D1	1 3/4"	2 1/4"	4 5/8"	pine	1	F
D2	1 7/8"	2 1/4"	2 1/8"	pine	1	F
D3	1 1/4"	2 1/4"	8 1/2"	pine	1	F
D4	1/2"	2 1/8"	2 3/8"	oak	2	R
D5	3/8"	3 1/4"	1 3/4"	pine	1	F
D6	1 3/4"	2 1/8"	2"	pine	1	R
D7	3/4"	2 1/4"	1 1/2"	oak	1	F
D8	3/8"	3 1/2"	4 7/8"	pine	1	F
D9	1/2"	1 1/2"	4 1/4"	oak	2	R
D10	1/2"	1 1/2"	4 1/4"	oak	2	R

Part	T	W	L	Material	Qty.	*
D11	1/4"	1/2"	1 1/4"	pine	6	F
D12	3/8"	1 7/8"	1 7/8"	maple	4	R
D13	3/8"	1 3/4"	1 3/4"	maple	4	R
D16	1/2"	5/8"	4 7/8"	pine	2	F
D17	1 1/2"	2 1/2"	8 3/4"	pine	2	R
D18	1/4"	2 1/4"	3"	maple	2	R
D19	3/8"	1 1/4"	3 1/4"	maple	1	R
D21	1 1/4"	4 1/16"	7 1/4"	pine	1	F
D22	5/8"	2 3/8"	6 1/4"	pine	1	R
D23	5/8"	5/8"	2 5/16"	maple	1	F

R = Rough size
F = Finished size

T = Thickness
W = Width
L = Length

Instructions:

R= Rough sizes, the material is cut oversized so you have ample room to apply the pattern on the surface. Sanding is not required at this point.

F = Finished Size: Cut and sand parts to finished size.

Full-Sized Patterns: Set One

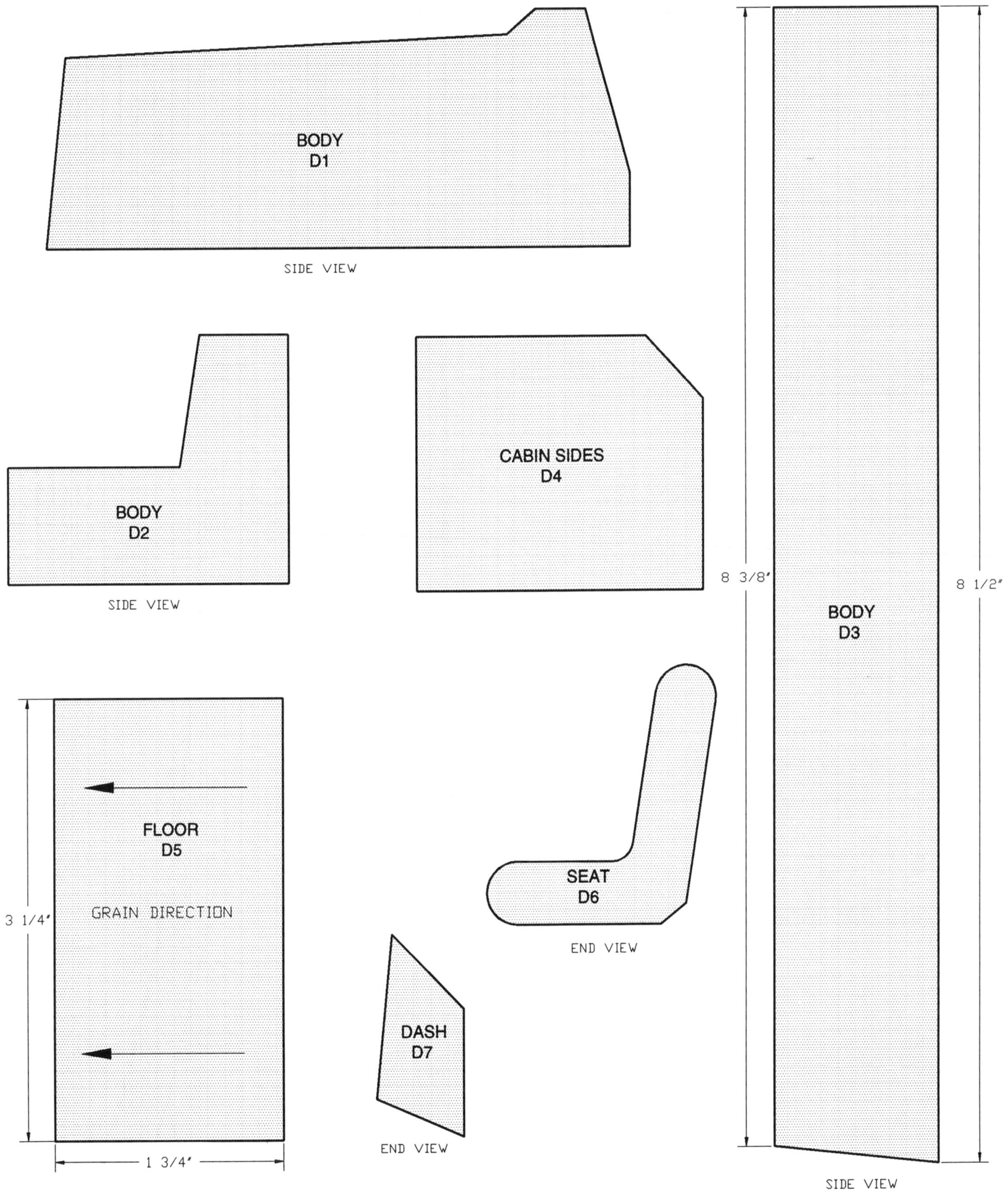

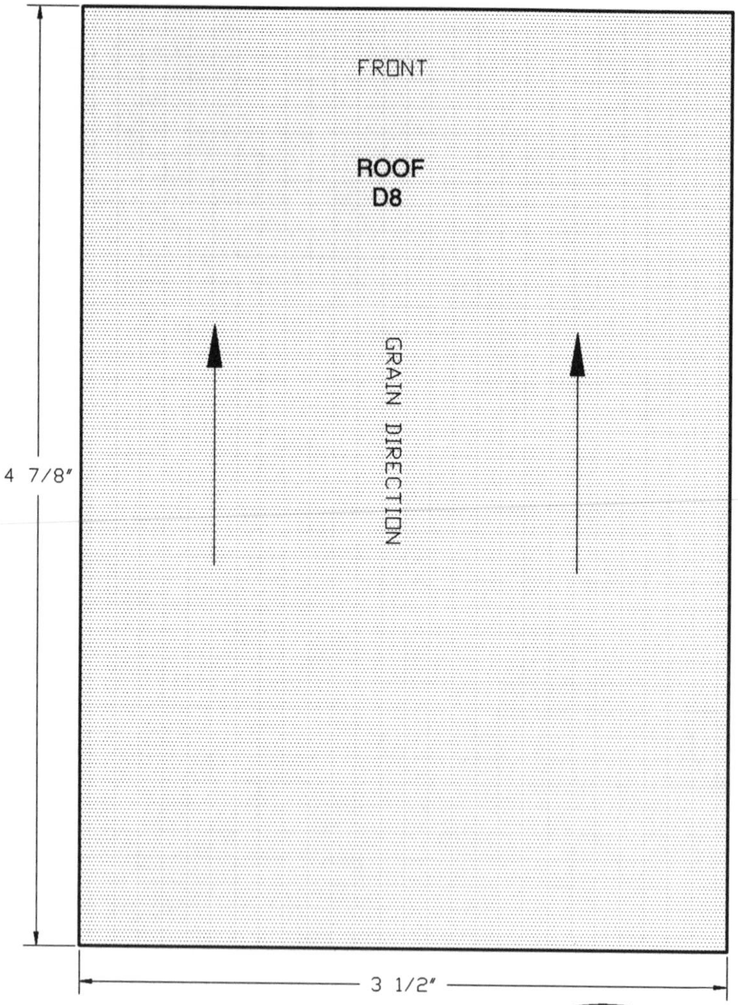

FRONT
ROOF
D8
GRAIN DIRECTION
4 7/8"
3 1/2"
CABIN FRAMES
FRONT
BACK
D9
D10
1 1/4"
D11
1/2"
TRACK SUPPORT BLOCKS
DRILL
23/64" DIA
WHEELS
D12
Ø1 11/16"
Ø1 9/16"
TRACK PULLEYS
D13

TRACK SUPPORTS
SIDE VIEW
D16

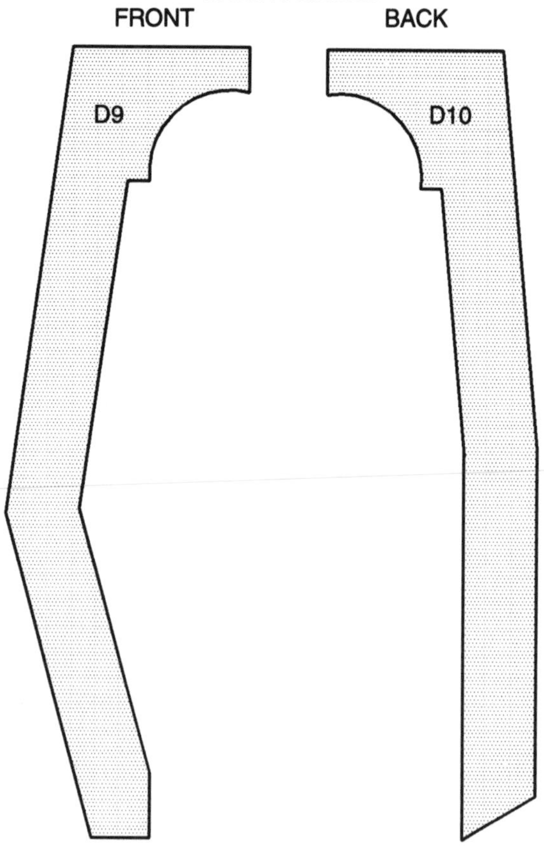

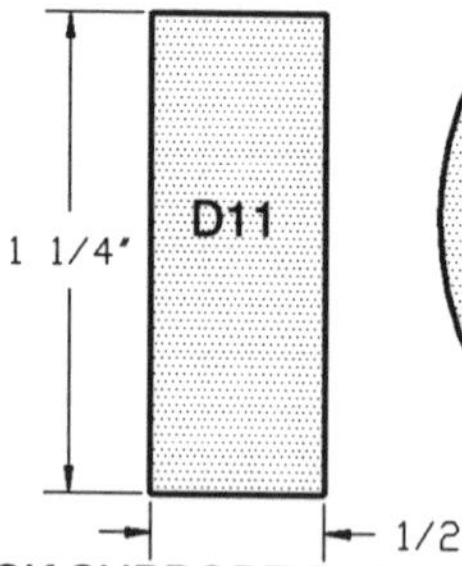

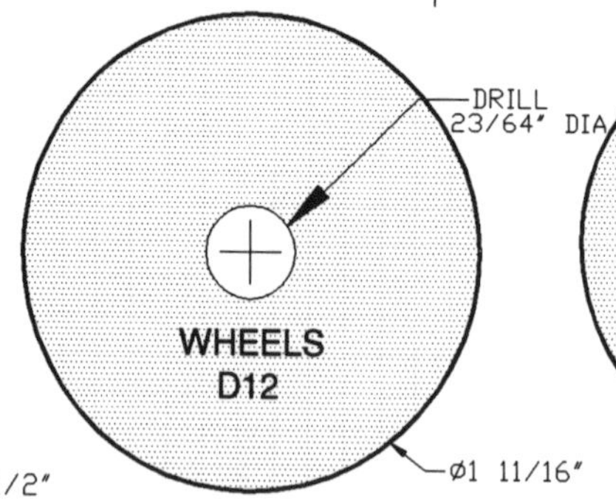

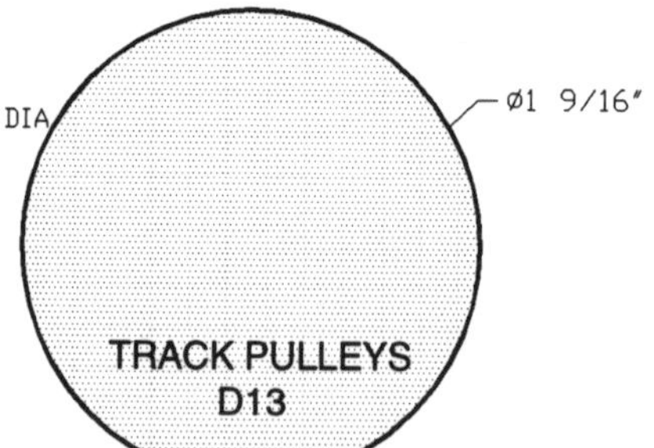

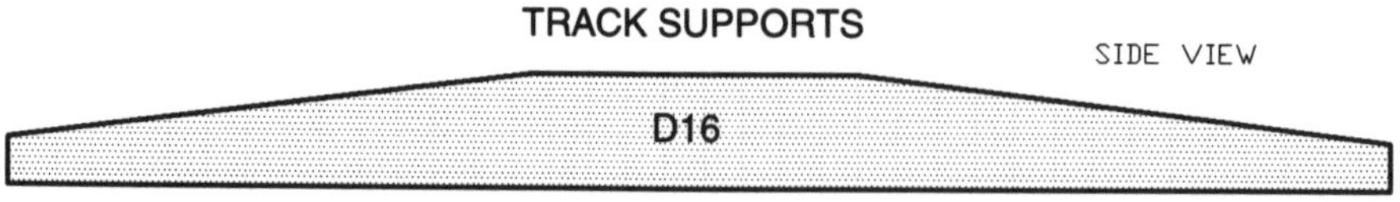

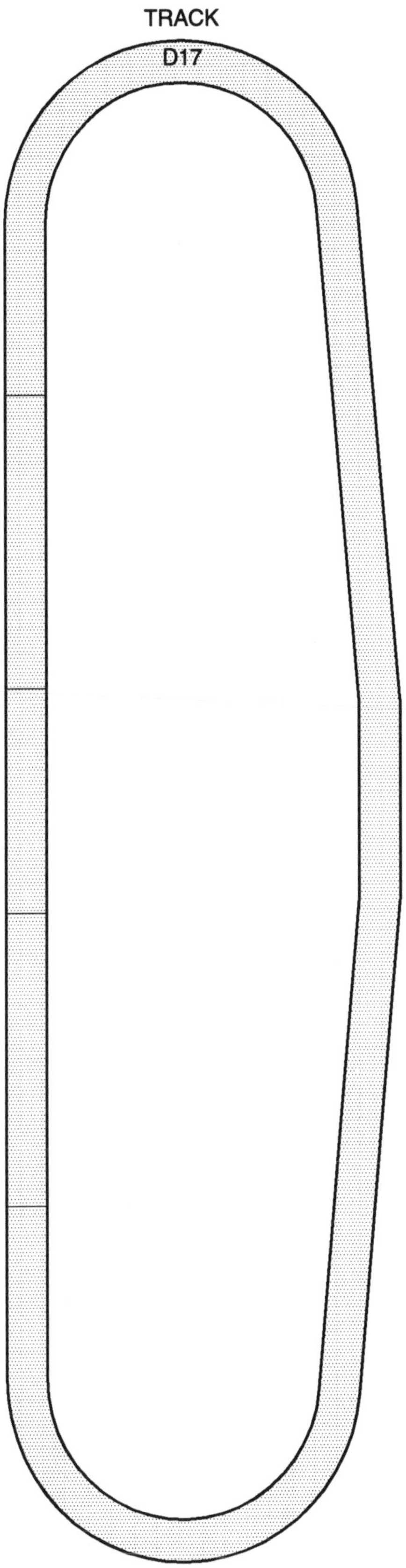
TRACK
D17

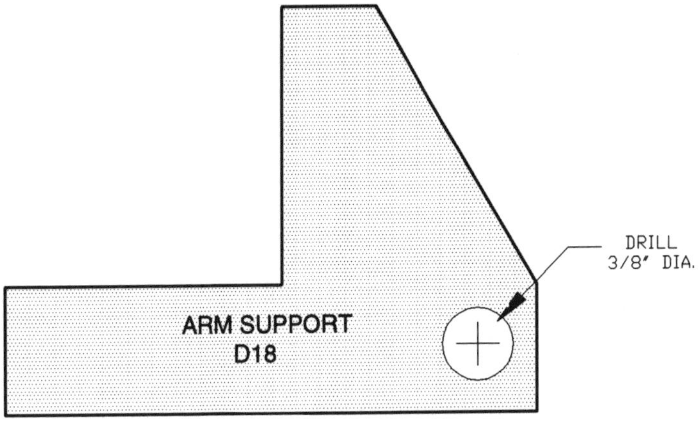
DRILL
3/8" DIA.
ARM SUPPORT
D18

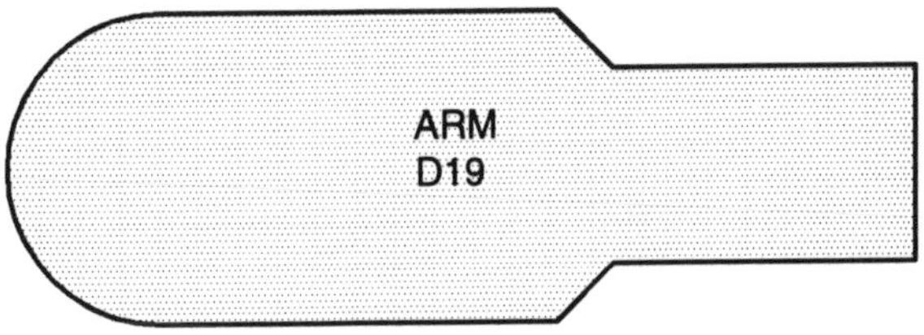
ARM
D19

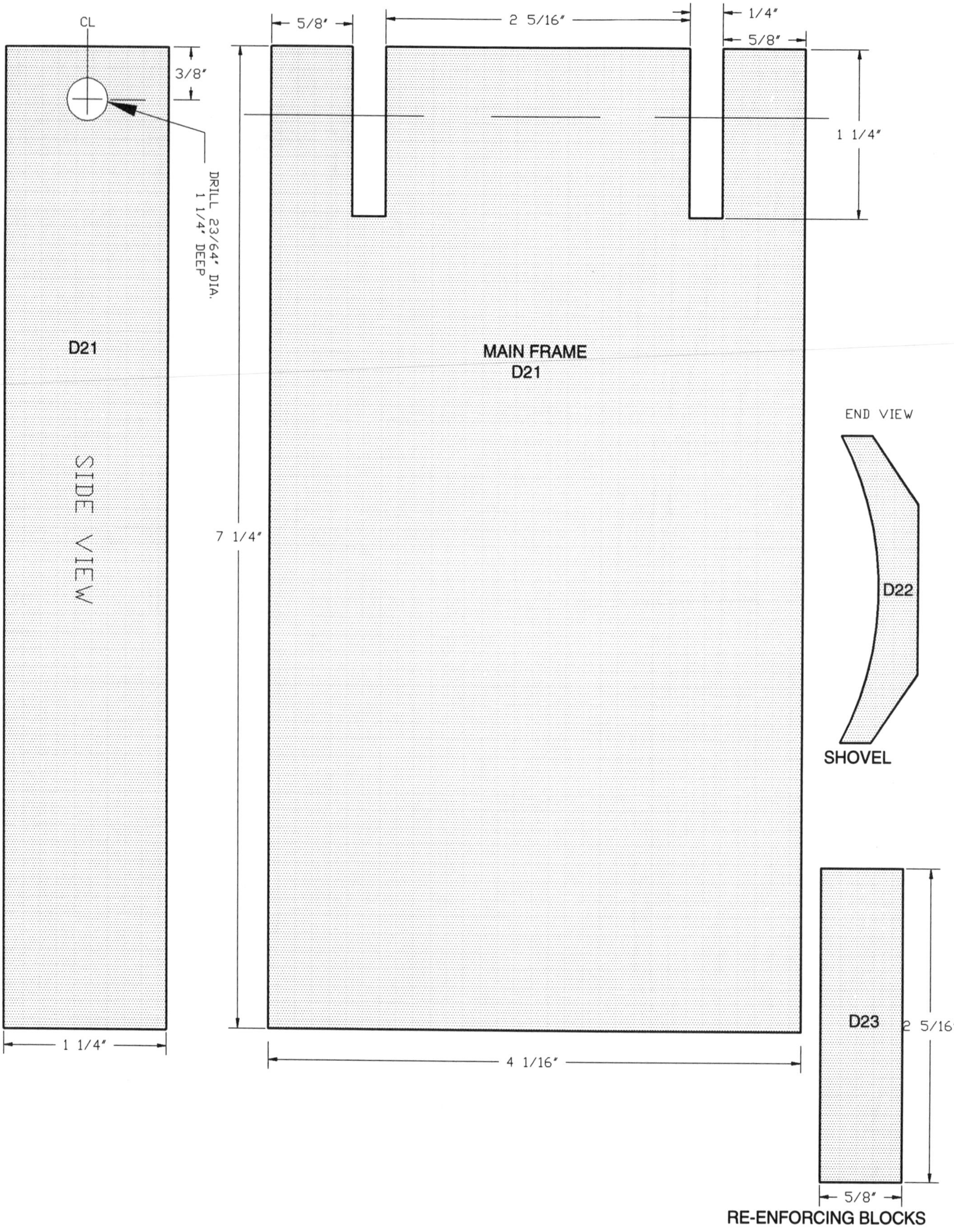

CL
3/8"
DRILL 23/64" DIA.
1 1/4" DEEP
D21
SIDE VIEW
1 1/4"
5/8"
2 5/16"
1/4"
5/8"
1 1/4"
7 1/4"
MAIN FRAME
D21
4 1/16"
END VIEW
D22
SHOVEL
D23
2 5/16"
5/8"
RE-ENFORCING BLOCKS

D24

WHEEL SHAFTS
(2 REQUIRED)

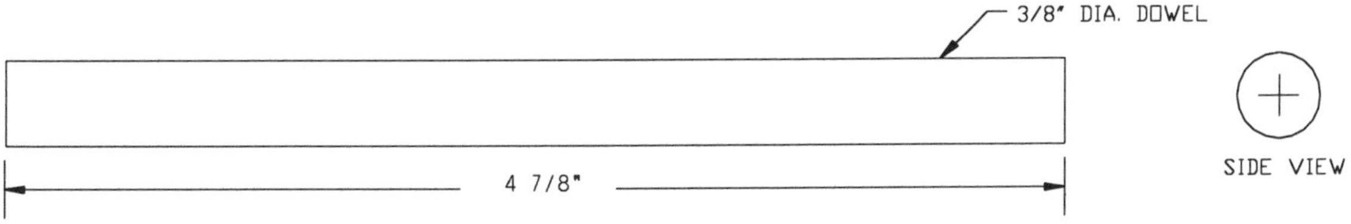

D26

PINS
(2 REQUIRED)

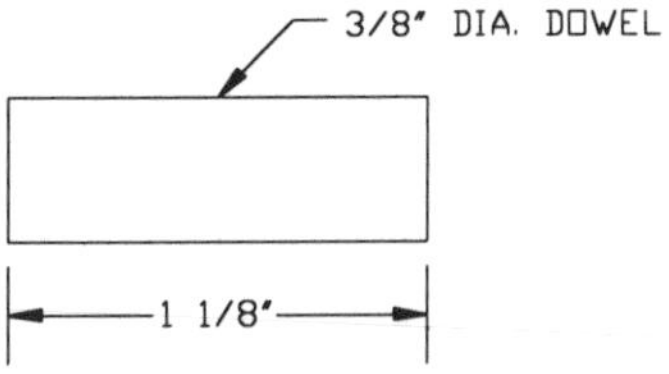

Additional Information - Dozer

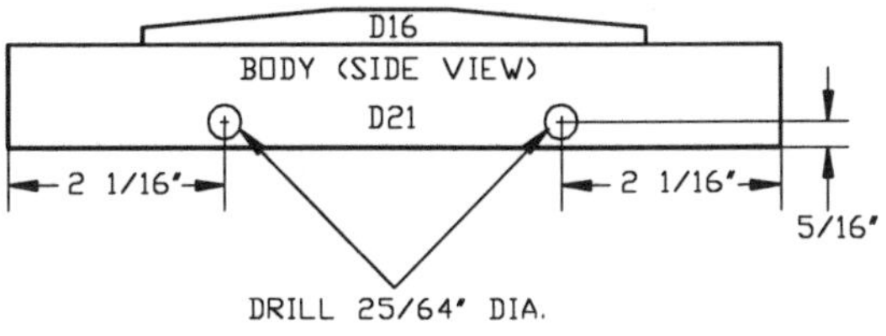

Drill holes into body D21, as shown.

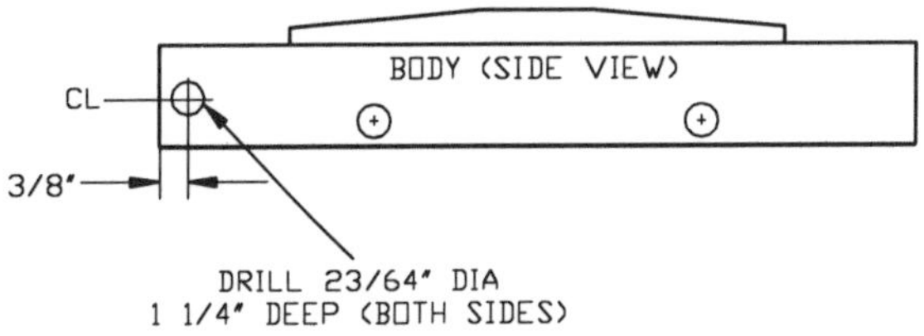

Drill 23/64" Dia. hole into body, as shown.

Dozer - Assembly Drawings

1

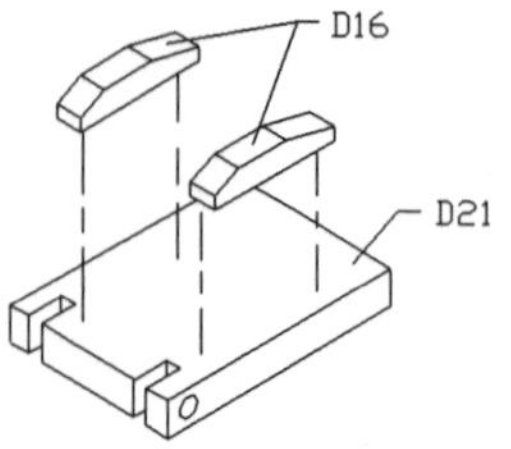

Glue track supports D16 onto frame D21, as shown.

2

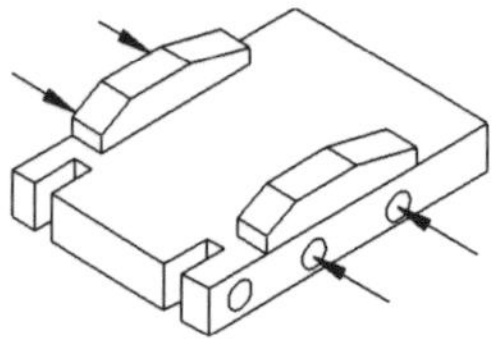

Trace and drill holes into frame D21 See page 29 for exact location.

3

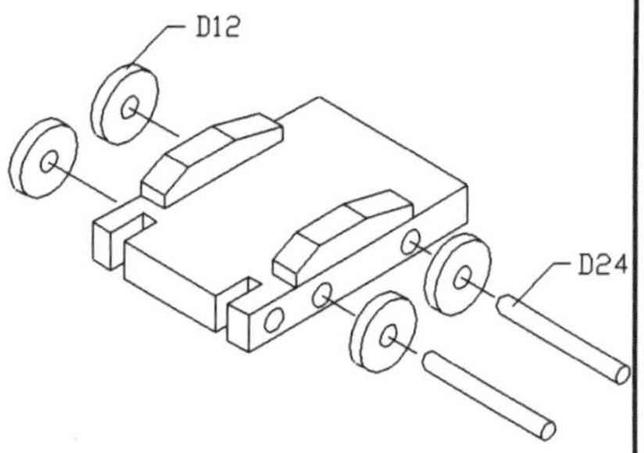

Insert wheel shafts D24 through wheels D12 and body.

4

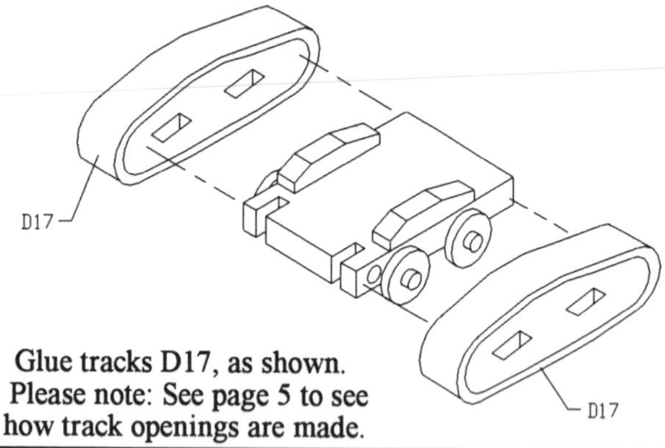

Glue tracks D17, as shown. Please note: See page 5 to see how track openings are made.

5

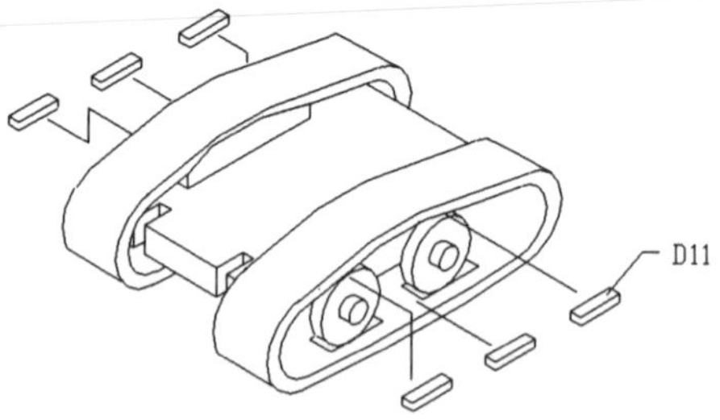

Glue track support blocks D11.

6

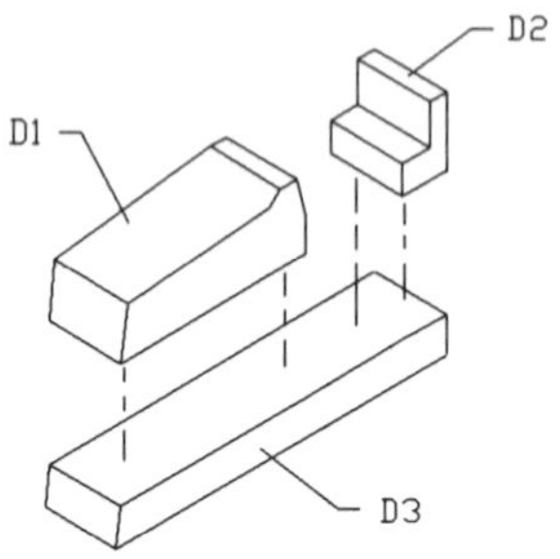

Glue parts, as shown.

7

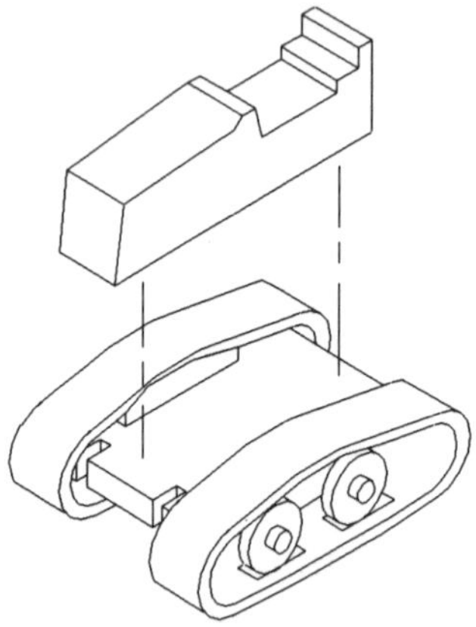

Glue body to track assembly.

8

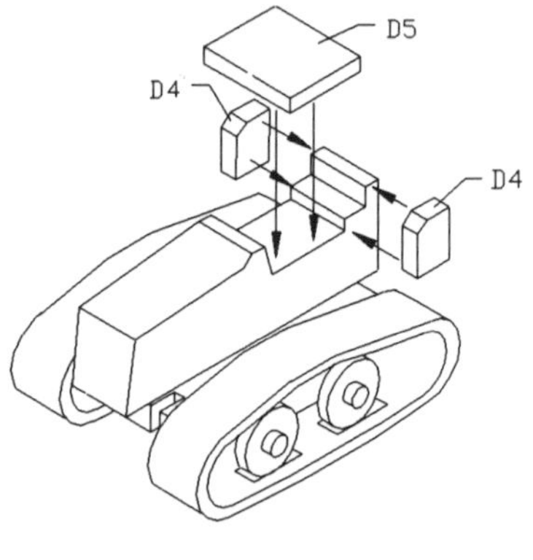

Glue floor D5 and sides D4.

9

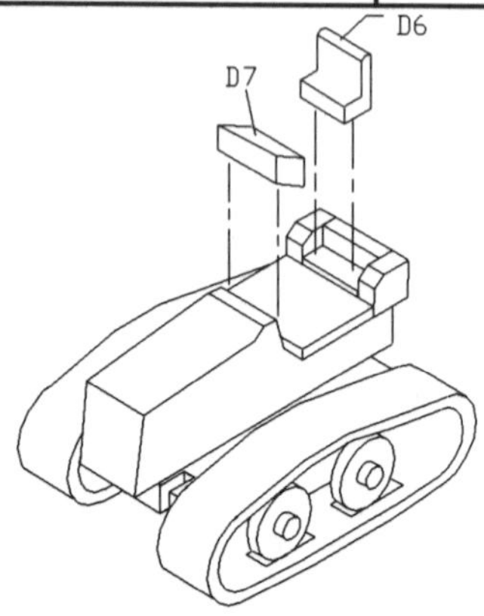

Glue seat D6 and dash D7.

10

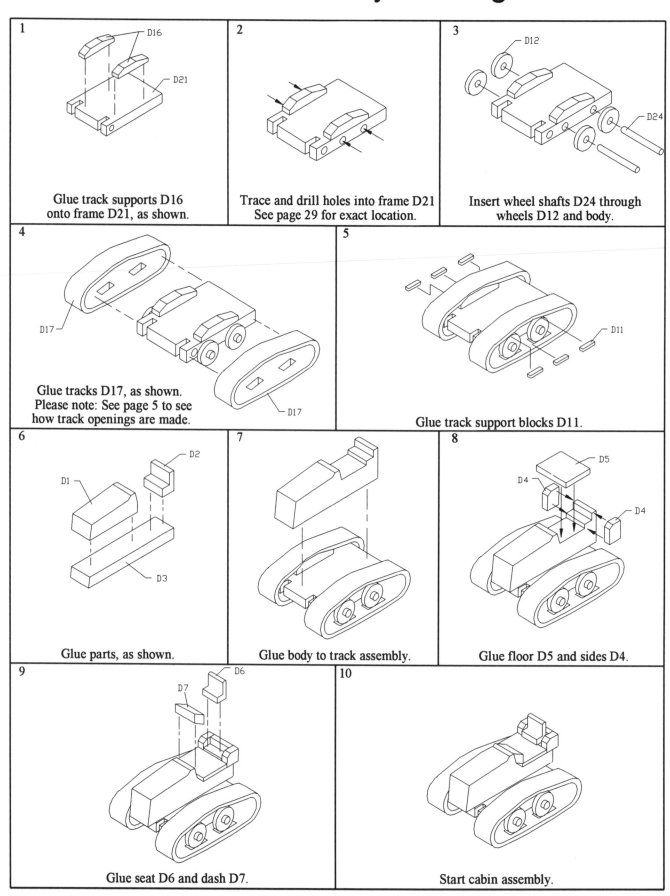

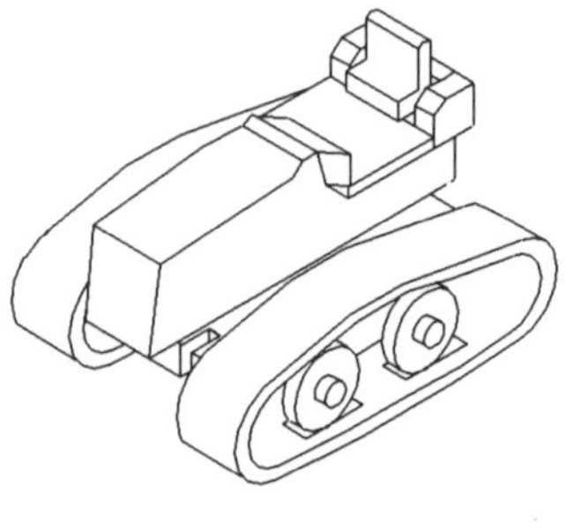

Start cabin assembly.

11 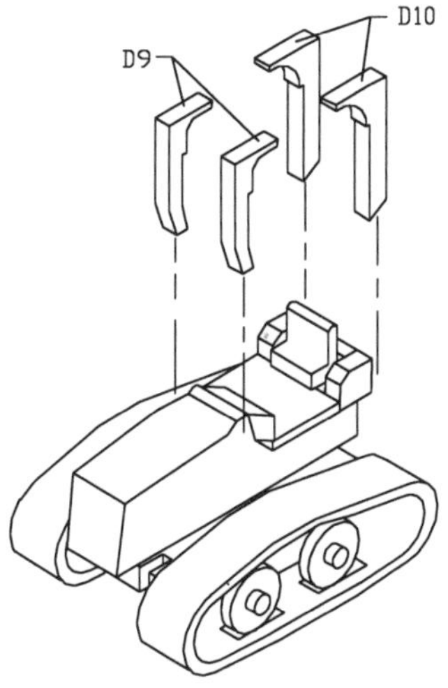 Glue cabin frames D9 and D10.	12 **IMPORTANT** Cabin must be completely assembled before glue dries so that you have time to make corrections, if necessary 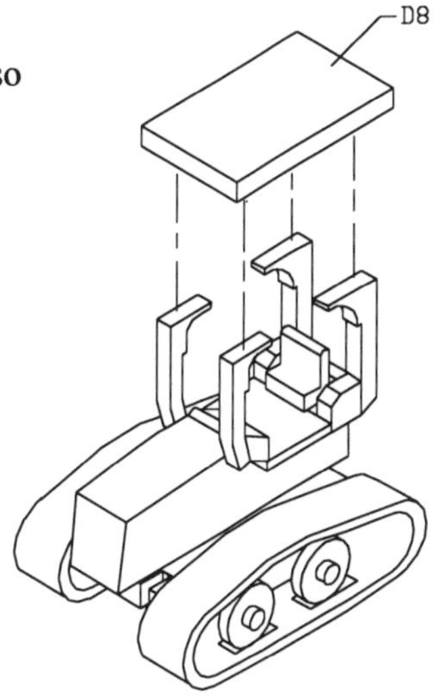 Glue roof D8, as shown.
13 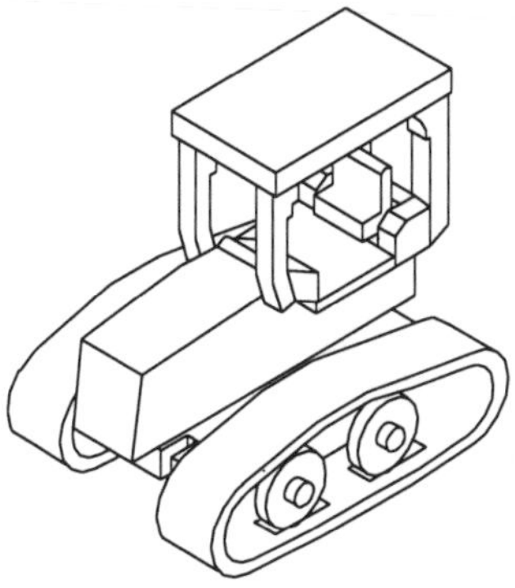 Start shovel assembly.	14 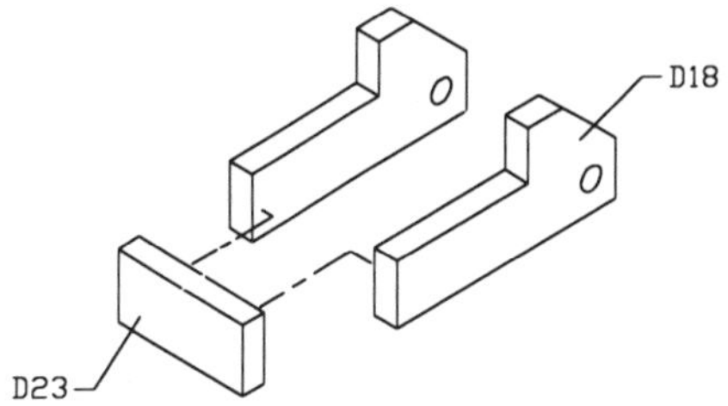 Glue re-enforcing block D23, as shown.
15 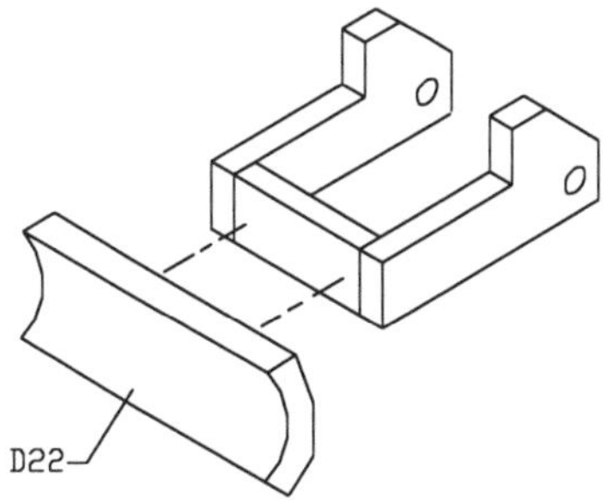 Glue shovel D22.	16 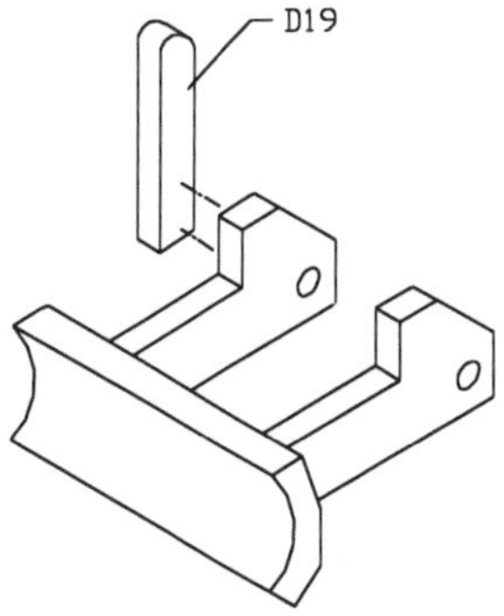 Glue arm D19, as shown.

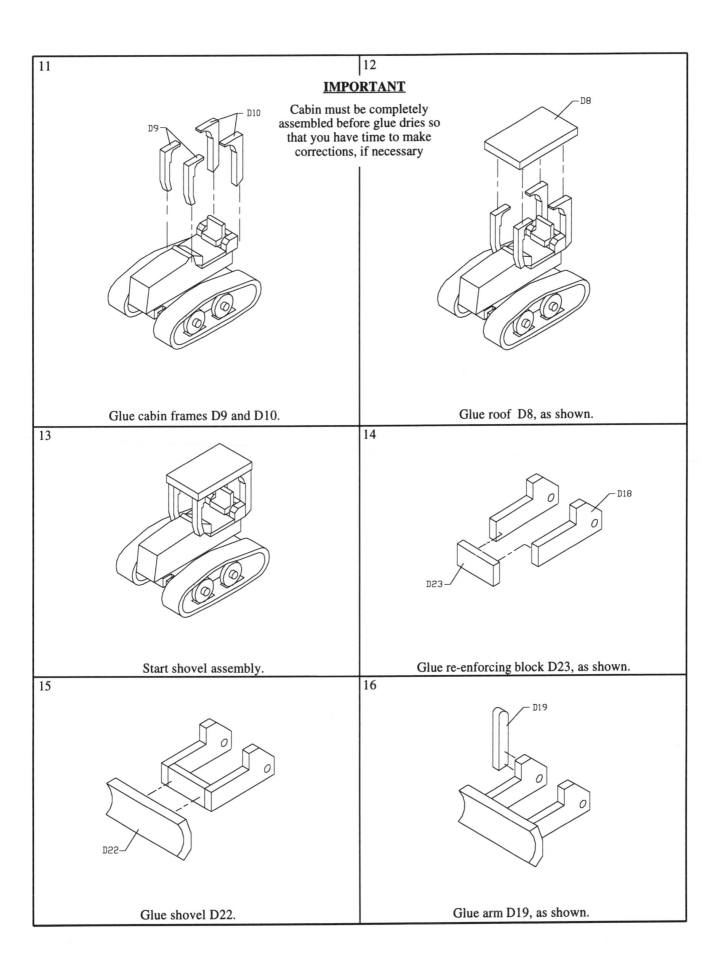

17

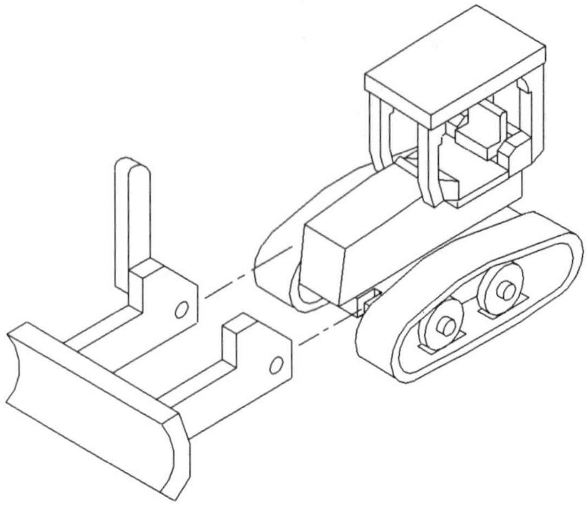

Slide shovel and arm assembly into body grooves.

18

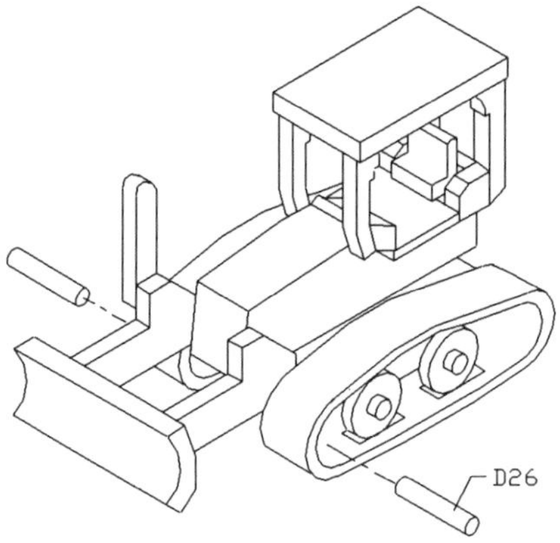

Insert pins D26, as shown.

19

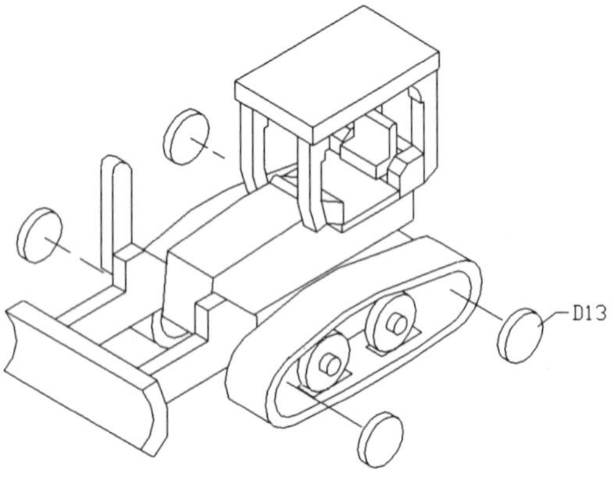

Glue track pulleys D13.

20

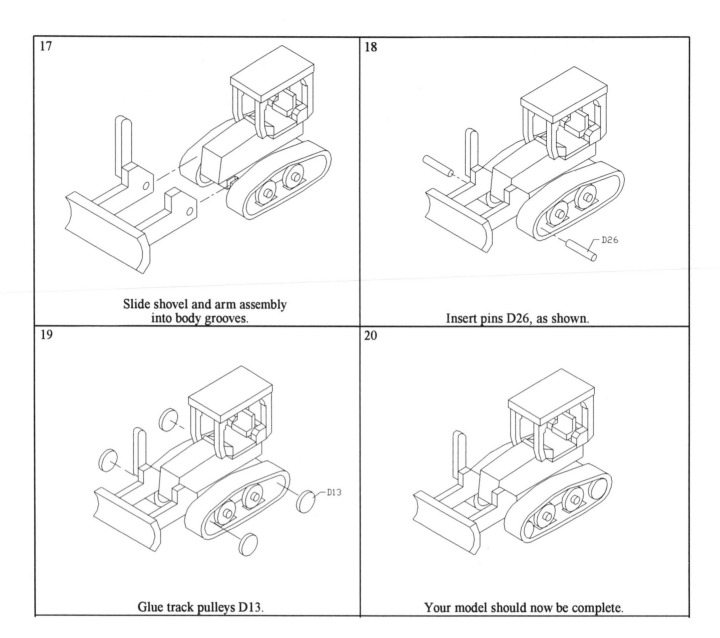

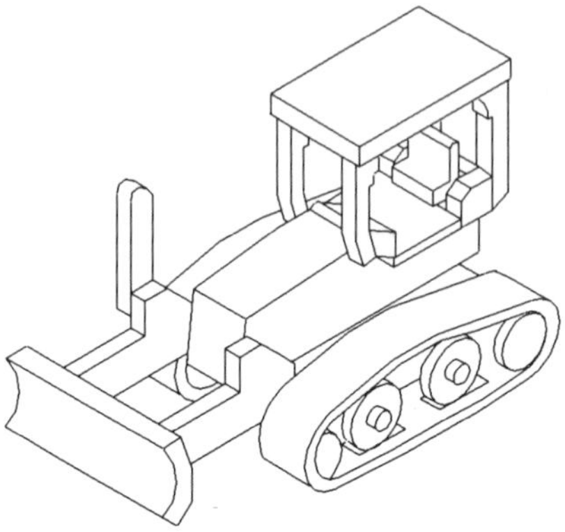

Your model should now be complete.

DOZER LOADER

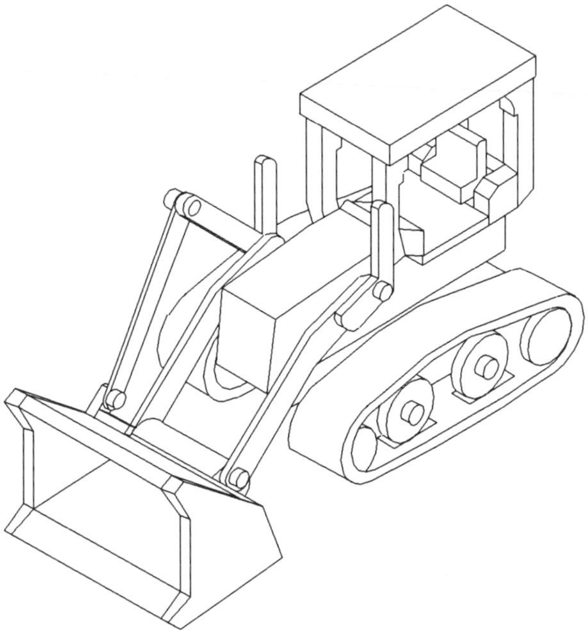

General Instructions - Dozer Loader

1- Start by cutting materials needed by following the list of materials, paying attention to the rough and finished size. **Identify the parts as they are cut.**

Please note: Different types of wood can be used for the various parts. It is suggested, however, that hard wood be used, since many of the parts would be much too fragile if using soft wood. We have used a combination of pine, maple and oak to give the models a nice contrast!

2- Remove the full-size patterns found in the appendix. Cut them out, leaving approximately 1/16" all around, and place on the proper piece of wood. Patterns can be secured to wood using either spray adhesive or rubber ciment. If using the latter, cut and sand the part first to finished size. If drilling is required, mark the hole by inserting a scriber or nail through the pattern into the wood. Remove the pattern before drilling.

You should have no trouble determining which surface to attach most of the patterns. Some parts, however, can be confusing since the pattern could fit on more than one surface. The drawings below indicate exactly which surface to attach the patterns for these parts.

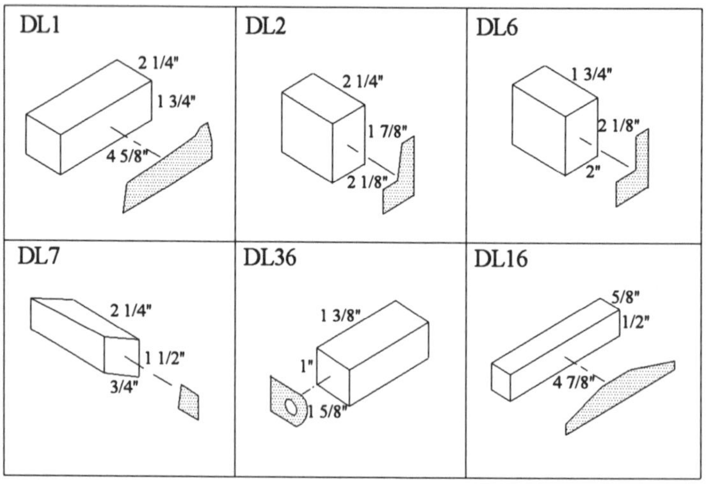

3- Look at the full-size drawing sheets to finish part DL3.

4- Part DL36 will need additional cuts and details, please refer to the additional information pages, to complete this part.

5- Using maple dowels, make all pins, shafts, etc.

6- Follow the assembly drawings to complete your model.

List of Materials - Dozer Loader

Part	T	W	L	Material	Qty.	*
DL1	1 3/4"	2 1/4"	4 5/8"	pine	1	F
DL2	1 7/8"	2 1/4"	2 1/8"	pine	1	F
DL3	1 1/4"	2 1/4"	8 1/2"	pine	1	F
DL4	1/2"	2 1/8"	2 3/8"	oak	2	R
DL5	3/8"	3 1/4"	1 3/4"	pine	1	F
DL6	1 3/4"	2 1/8"	2"	pine	1	R
DL7	3/4"	2 1/4"	1 1/2"	oak	1	F
DL8	3/8"	3 1/2"	4 7/8"	pine	1	F
DL9	1/2"	1/2"	4 1/4"	oak	2	R
DL10	1/2"	1/2"	4 1/4"	oak	2	R
DL11	1/4"	1/2"	1 1/4"	pine	6	F
DL12	3/8"	1 7/8"	1 7/8"	maple	4	R
DL13	3/8"	1 3/4"	1 3/4"	maple	4	R
DL16	3/8"	5/8"	4 7/8"	pine	2	F

Part	T	W	L	Material	Qty.	*
DL17	1 1/2"	2 1/2"	8 3/4"	pine	2	R
DL18	1"	1 1/8"	2 1/4"	maple	1	F
DL19	3/8"	3 3/8"	3 1/2"	pine	2	R
DL20	3/8"	2"	5"	maple	1	R
DL21	3/8"	1 1/8"	3"	maple	1	R
DL22	3/8"	3"	3 1/8"	maple	1	R
DL23	3/8"	3 3/8"	5 1/2"	pine	1	F
DL24	3/8"	2 5/8"	5 1/2"	pine	1	F
DL25	1/4"	2 1/8"	6 3/8"	pine	1	R
DL26	3/8"	1 3/8"	6 1/8"	maple	2	R
DL28	1 1/4"	4 1/16"	7 1/4"	pine	1	F
DL36	1"	1 3/8"	1 5/8"	maple	1	F
DL37	3/8"	1"	6 1/4"	maple	1	R

T = Thickness
W = Width
L = Length

R = Rough size
F = Finished size

Instructions:

R= Rough sizes, the material is cut oversized so you have ample room to apply the pattern on the surface. Sanding is not required at this point.

F = Finished Size: Cut and sand parts to finished size.

Full-Sized Patterns: Set One

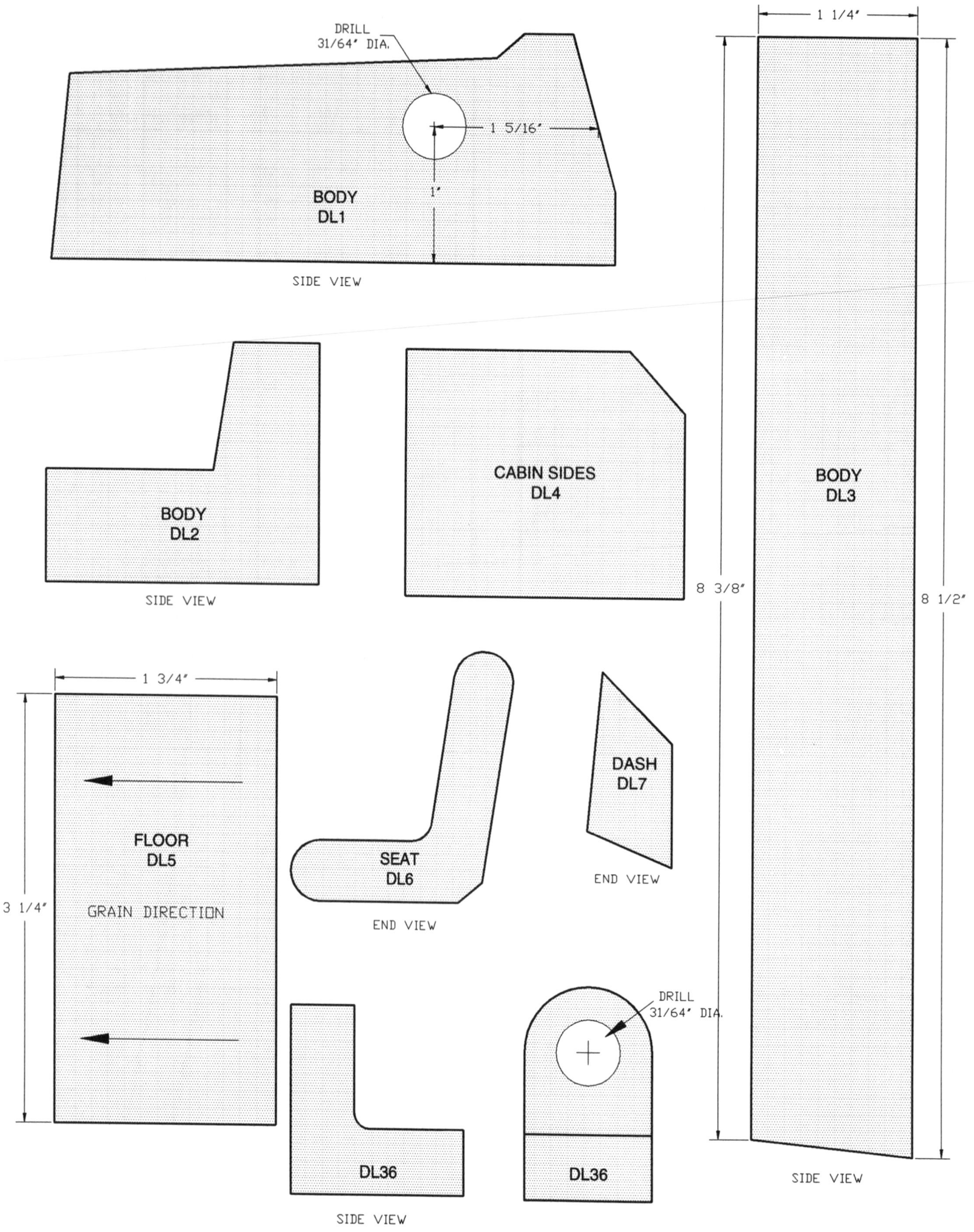

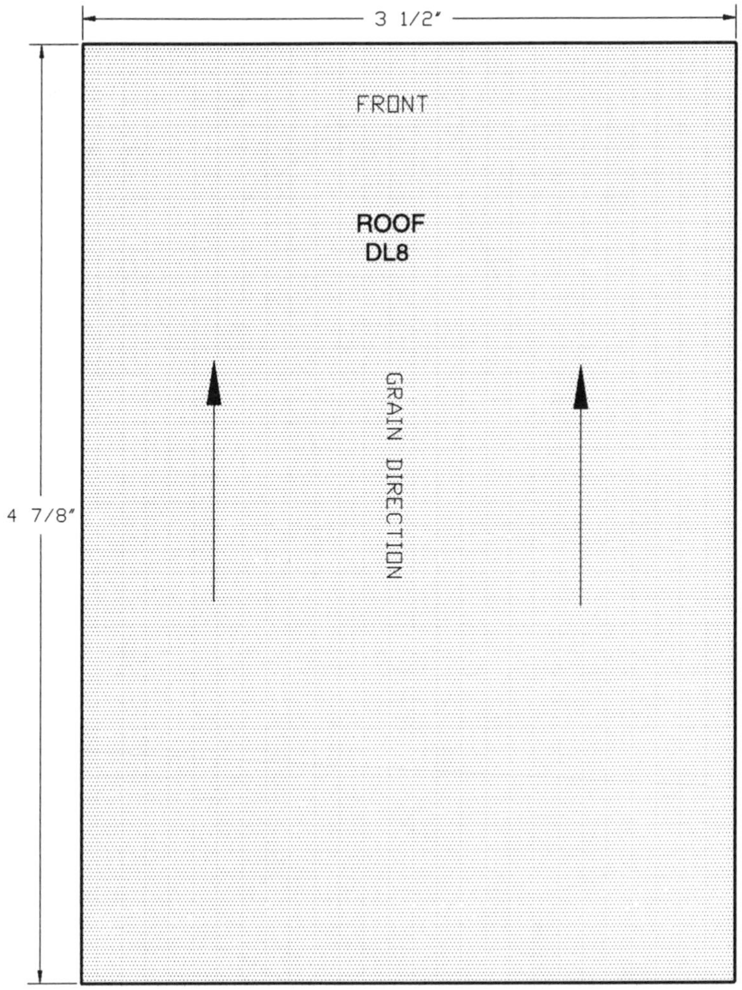

CABIN FRAMES

FRONT

BACK

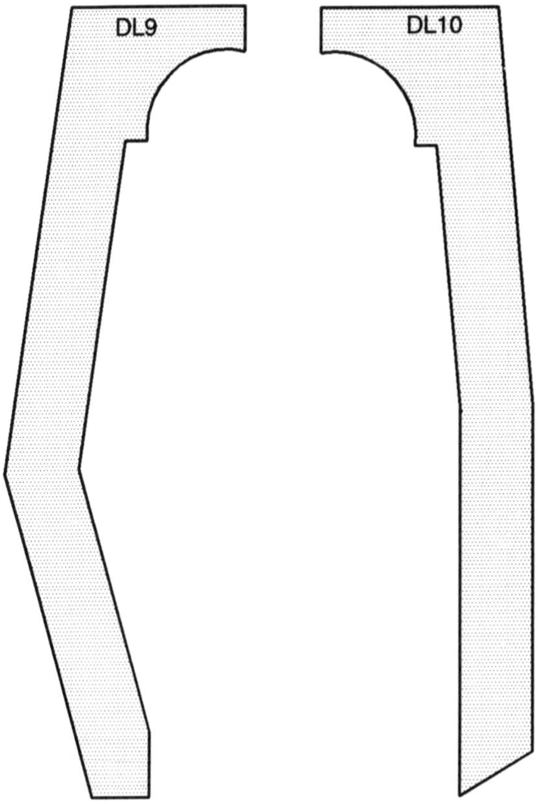

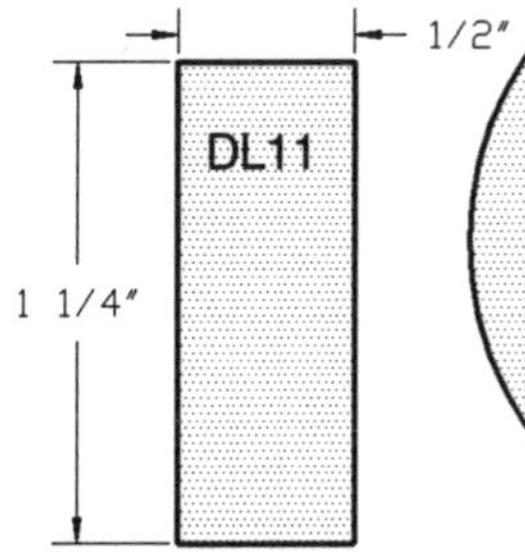

TRACK SUPPORT BLOCKS

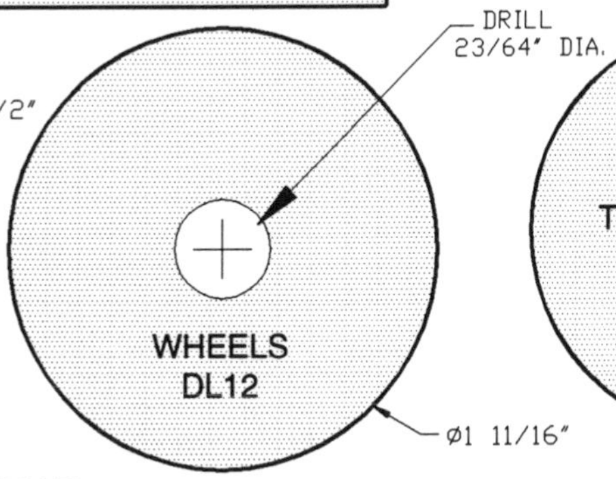

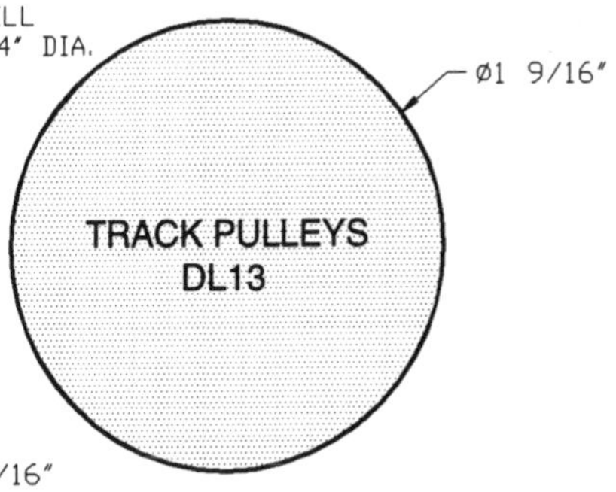

TRACK SUPPORTS

SIDE VIEW

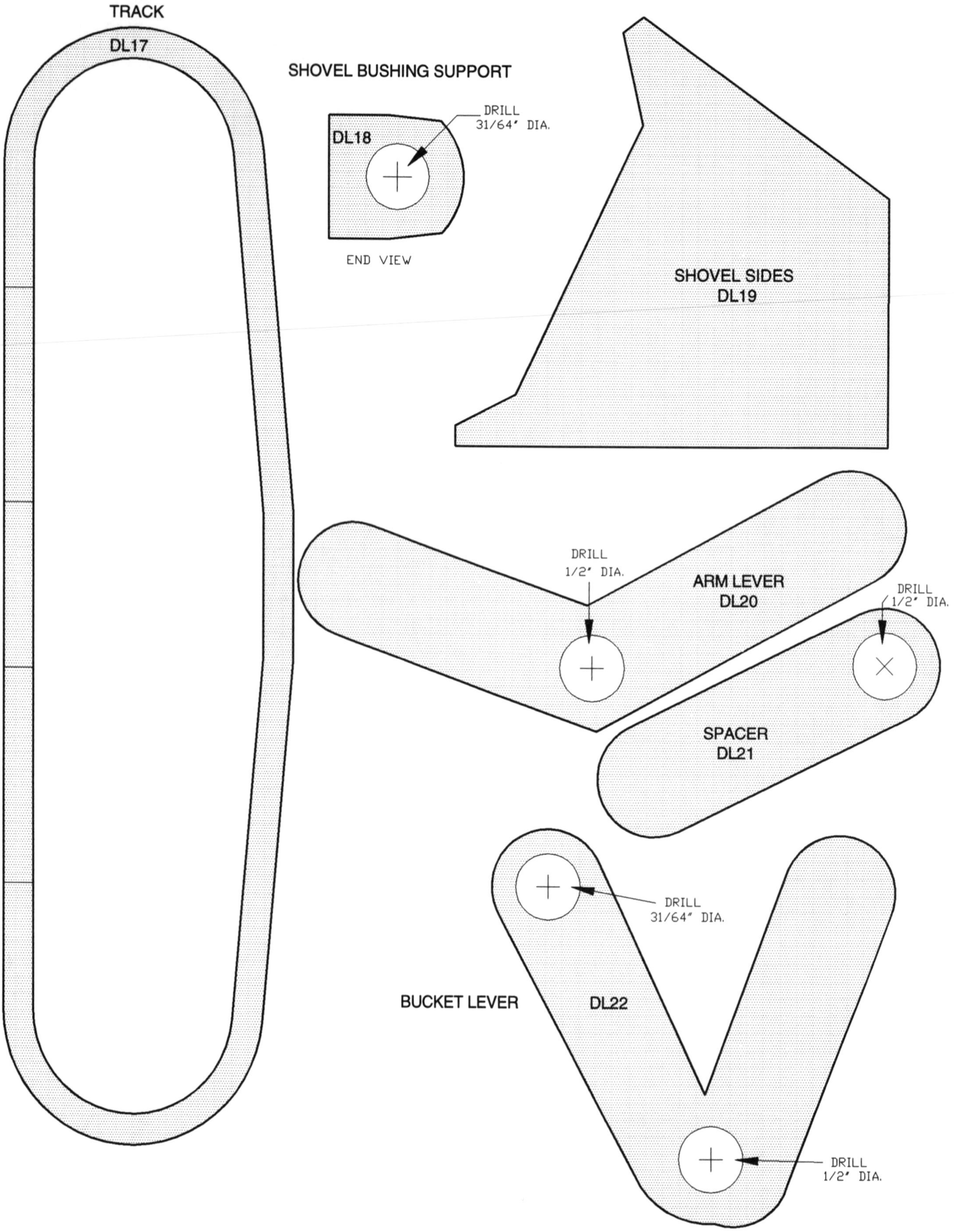
TRACK
DL17
SHOVEL BUSHING SUPPORT
DL18
DRILL
31/64" DIA.
END VIEW
SHOVEL SIDES
DL19
ARM LEVER
DL20
DRILL
1/2" DIA.
DRILL
1/2" DIA.
SPACER
DL21
DRILL
31/64" DIA.
BUCKET LEVER
DL22
DRILL
1/2" DIA.

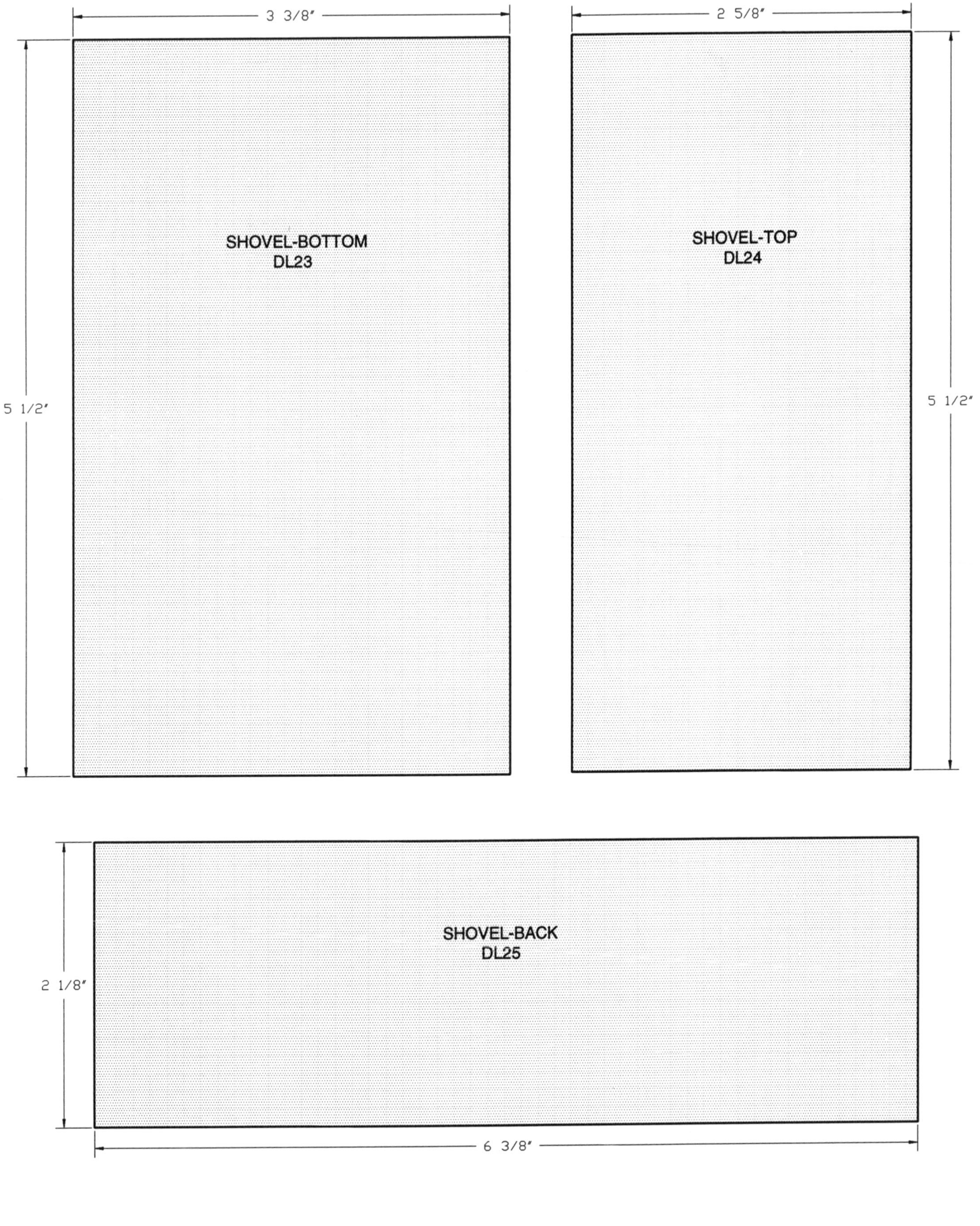
3 3/8"
2 5/8"
SHOVEL-BOTTOM
DL23
SHOVEL-TOP
DL24
5 1/2"
5 1/2"
SHOVEL-BACK
DL25
2 1/8"
6 3/8"

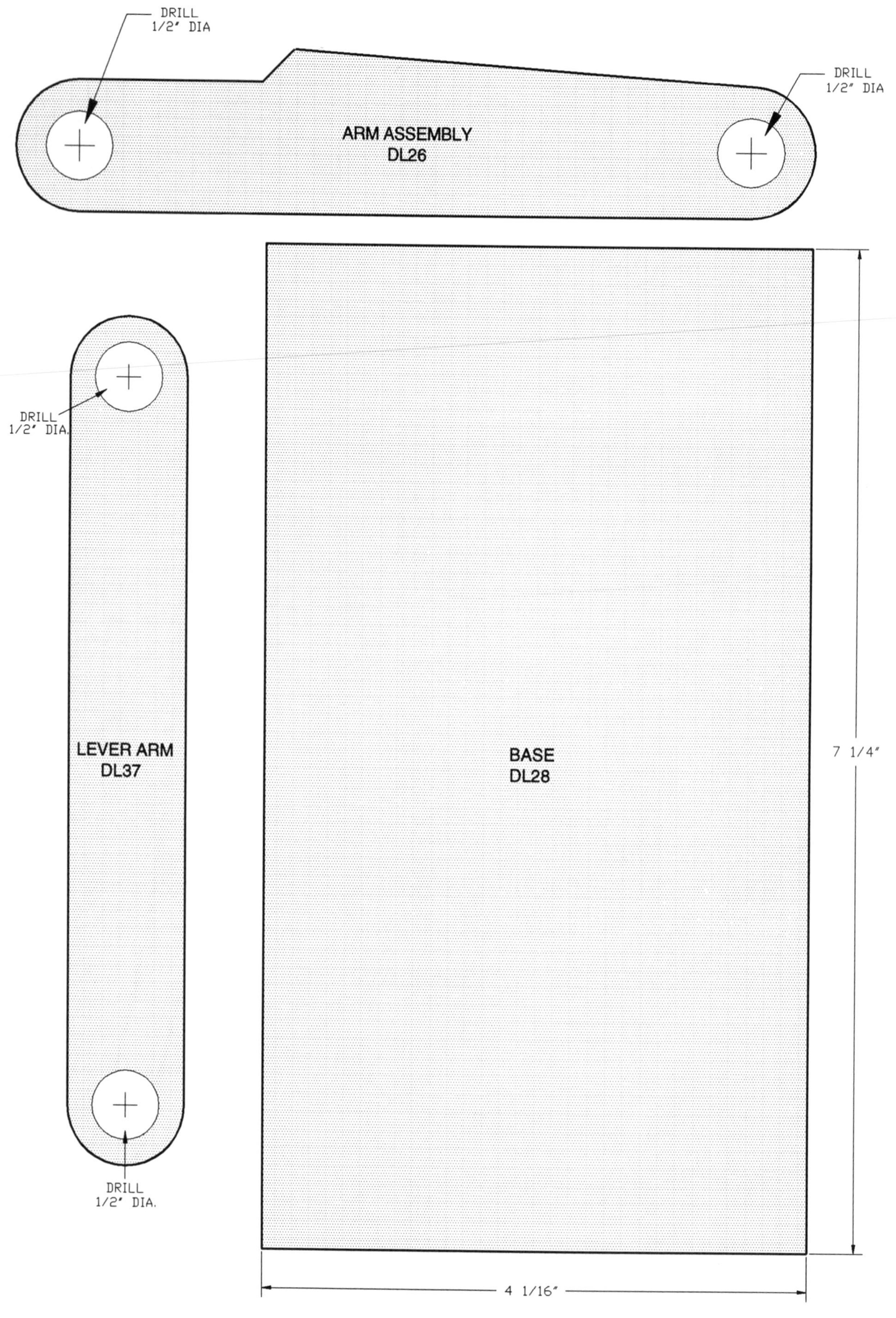
DRILL
1/2" DIA
ARM ASSEMBLY
DL26
DRILL
1/2" DIA
DRILL
1/2" DIA.
LEVER ARM
DL37
DRILL
1/2" DIA.
BASE
DL28
7 1/4"
4 1/16"

WHEEL SHAFTS
(2 REQUIRED)

DL29

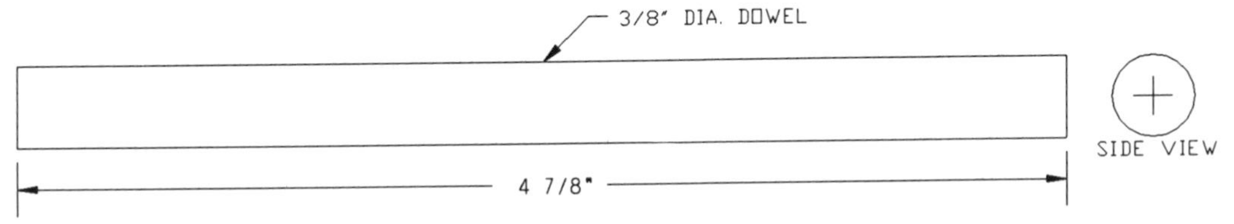

PINS
(2 REQUIRED)

DL32

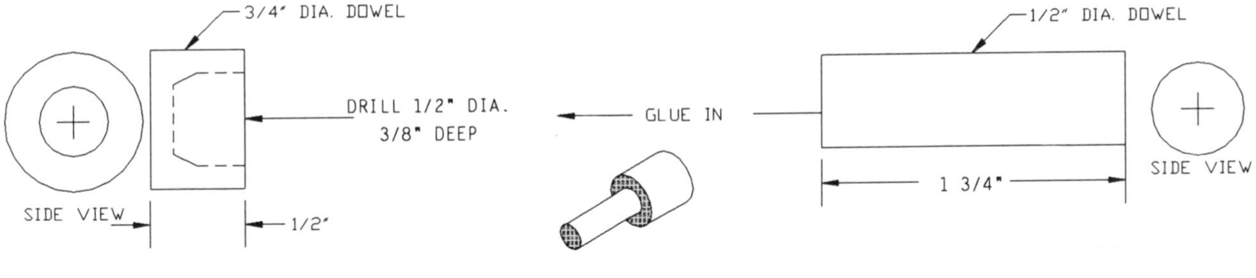

ASSEMBLED PART

PINS
(2 REQUIRED)

DL33

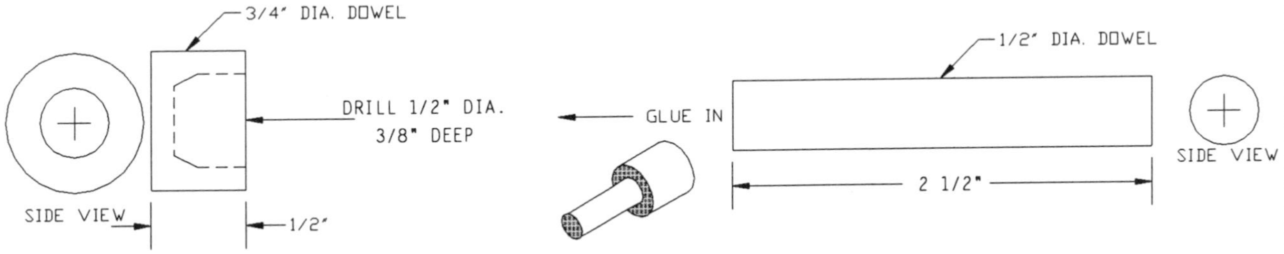

ASSEMBLED PART

LOCKING PINS
(2 REQUIRED)

DL34

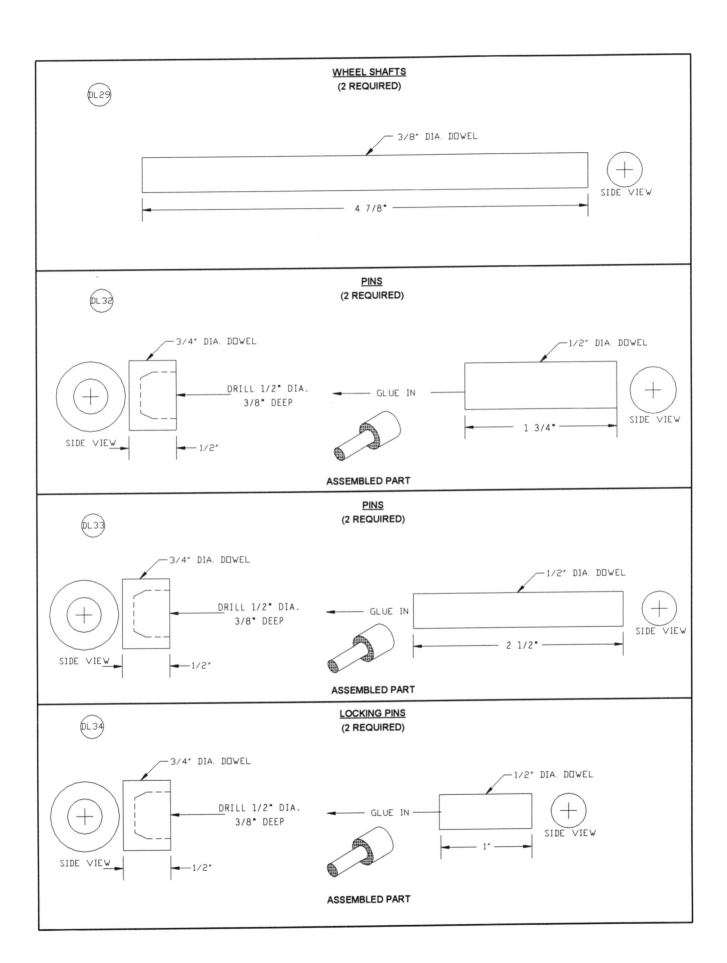

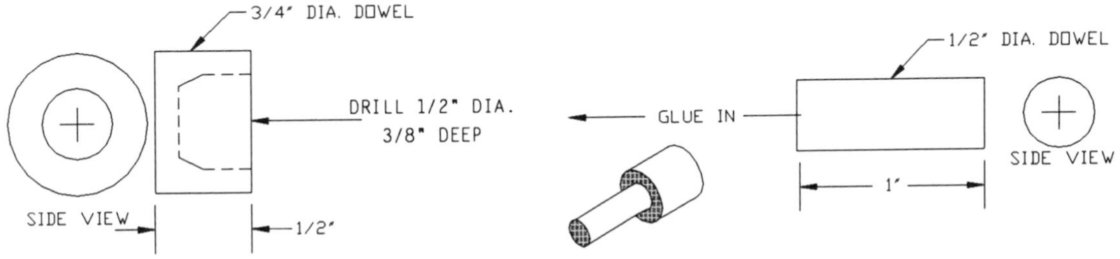

ASSEMBLED PART

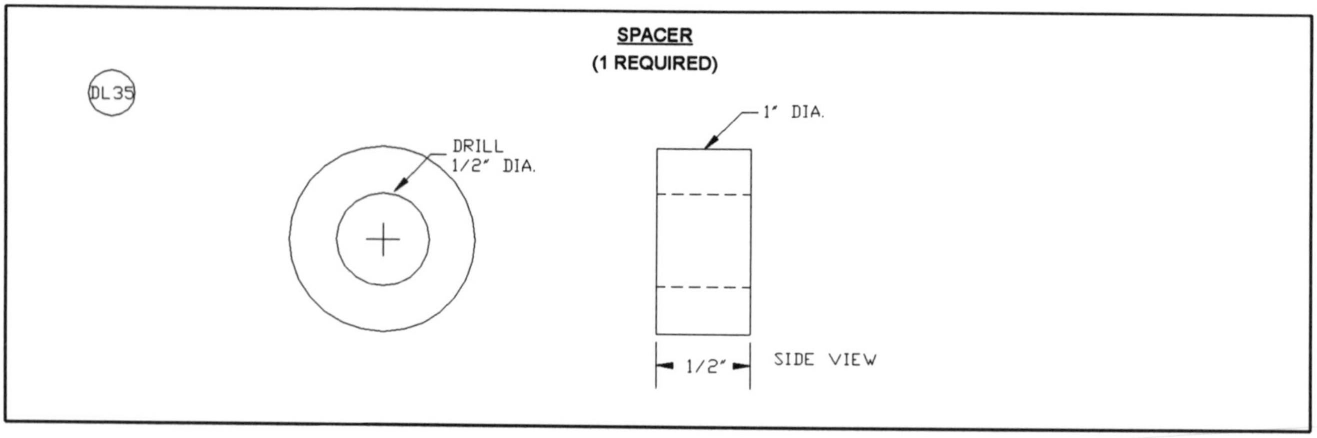

Additional Information - Dozer Loader

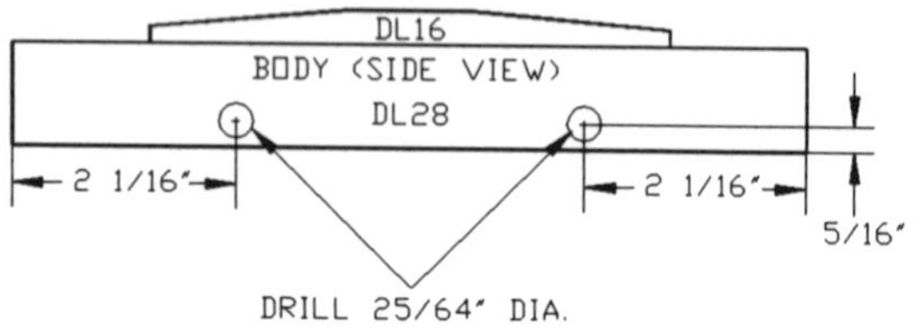

Drill holes into body DL28, as shown.

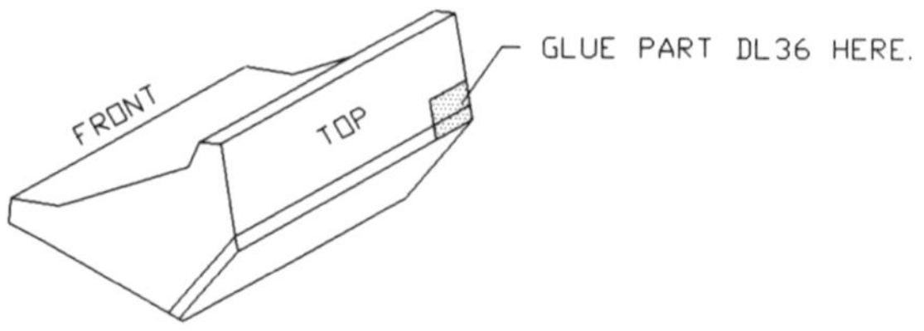

Where to glue part DL36 onto shovel assembly.

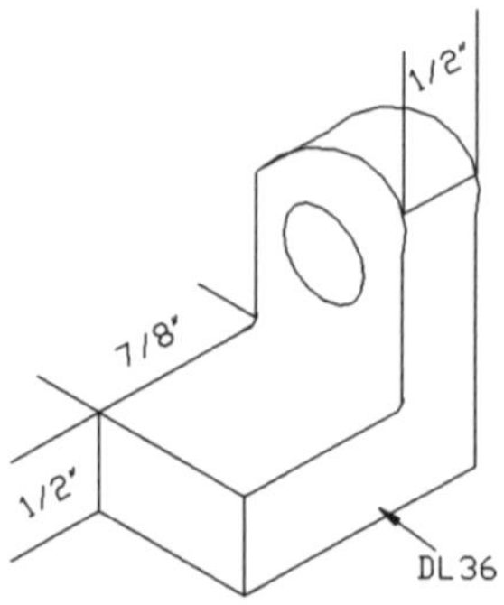

Part DL36 completed.

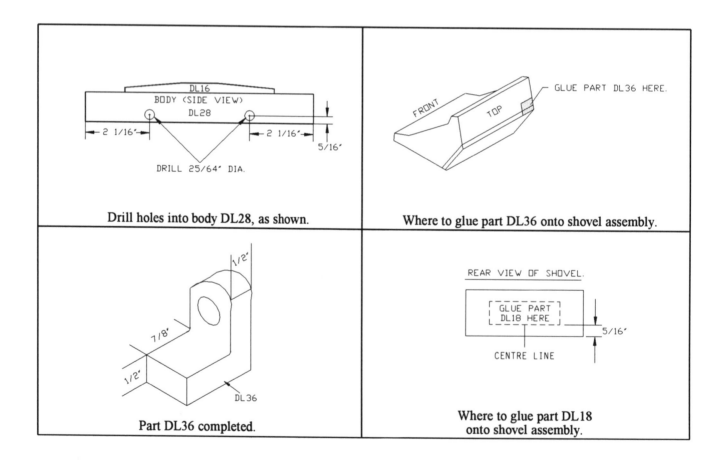

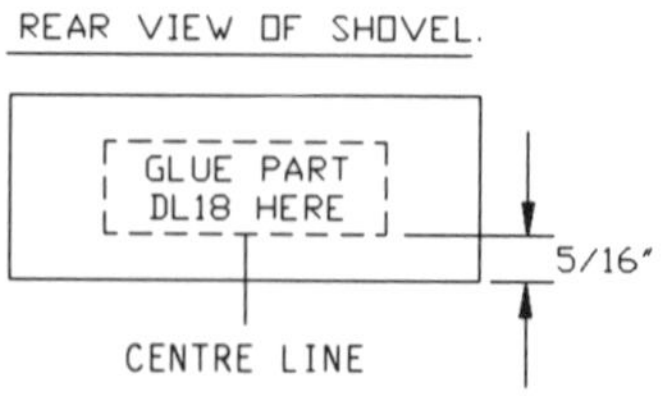

Where to glue part DL18
onto shovel assembly.

Dozer Loader - Assembly Drawings

1

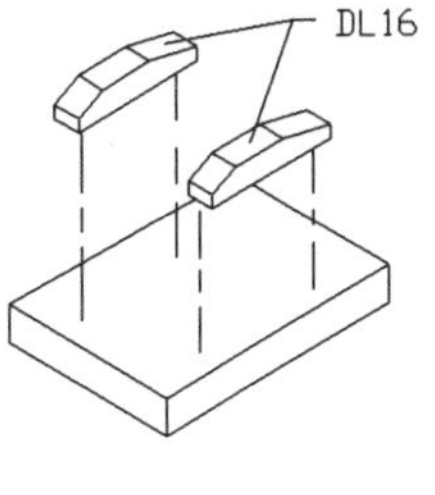

Glue track supports DL16 onto base DL28.

2

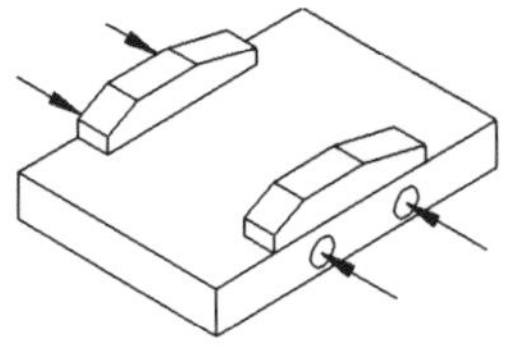

Drill 4 holes using method shown on page 42.

3

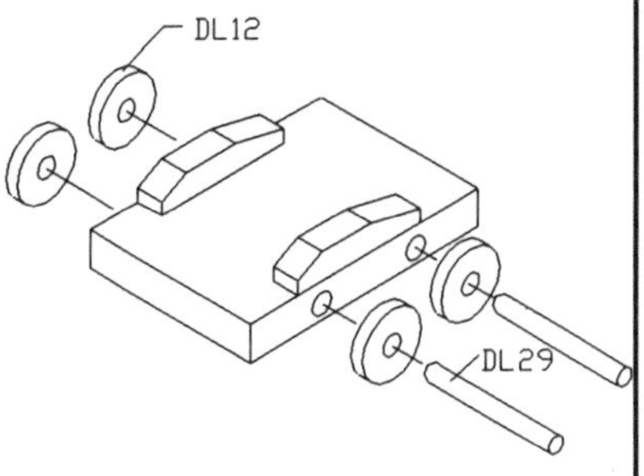

Insert wheel shafts DL29 through wheels DL12 and body.

4

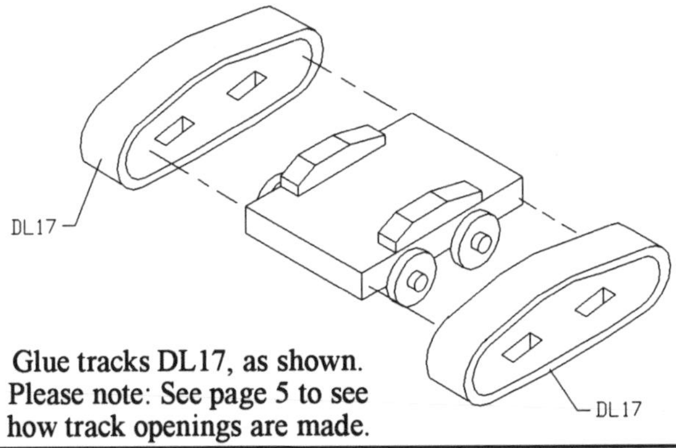

Glue tracks DL17, as shown. Please note: See page 5 to see how track openings are made.

5

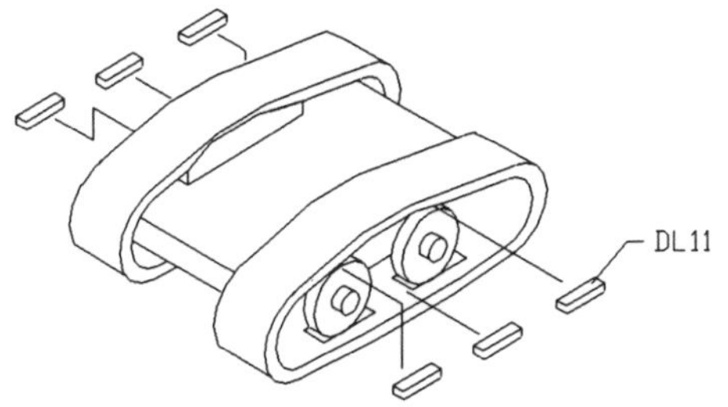

Glue track support blocks DL11.

6

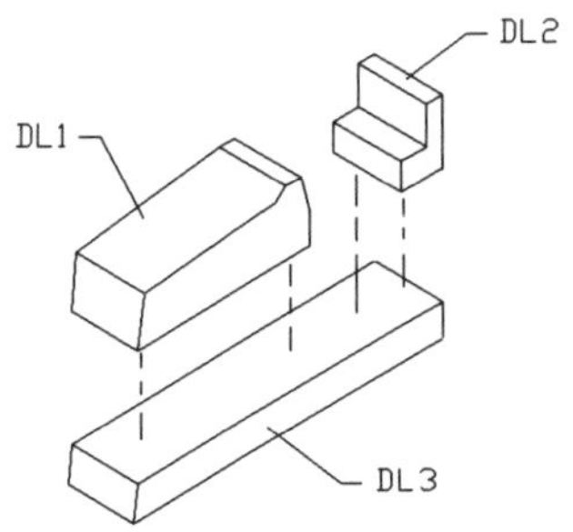

Glue main body, as shown.

7

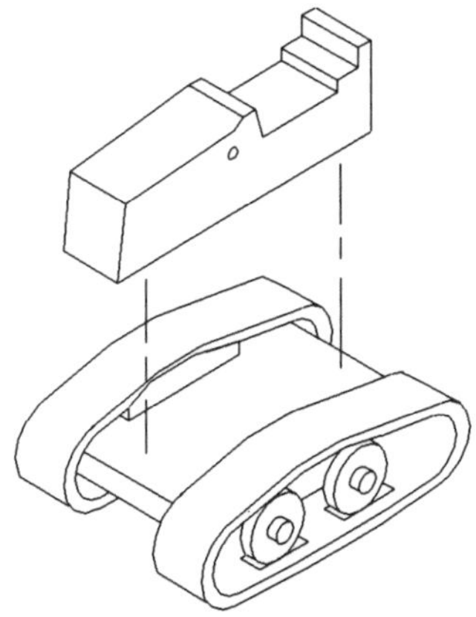

Glue body to track assembly.

8

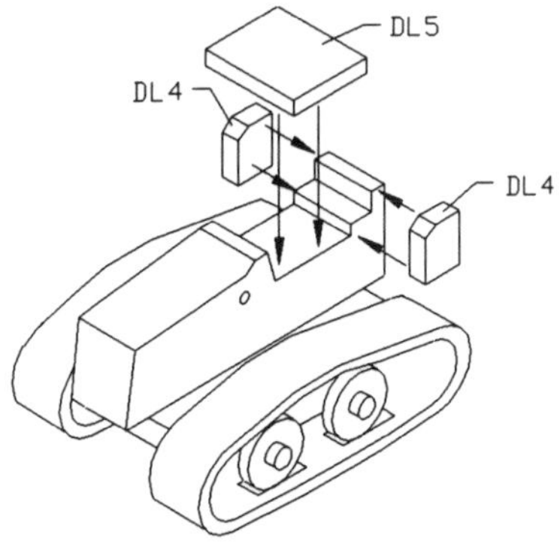

Glue floor DL5 and sides DL4.

9

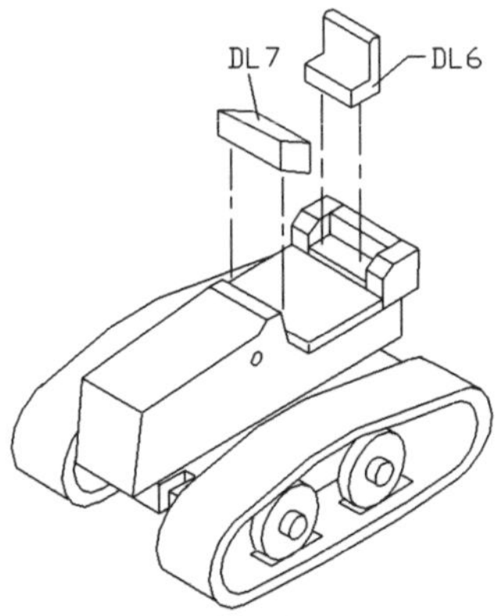

Glue seat DL6 and dash DL7.

10

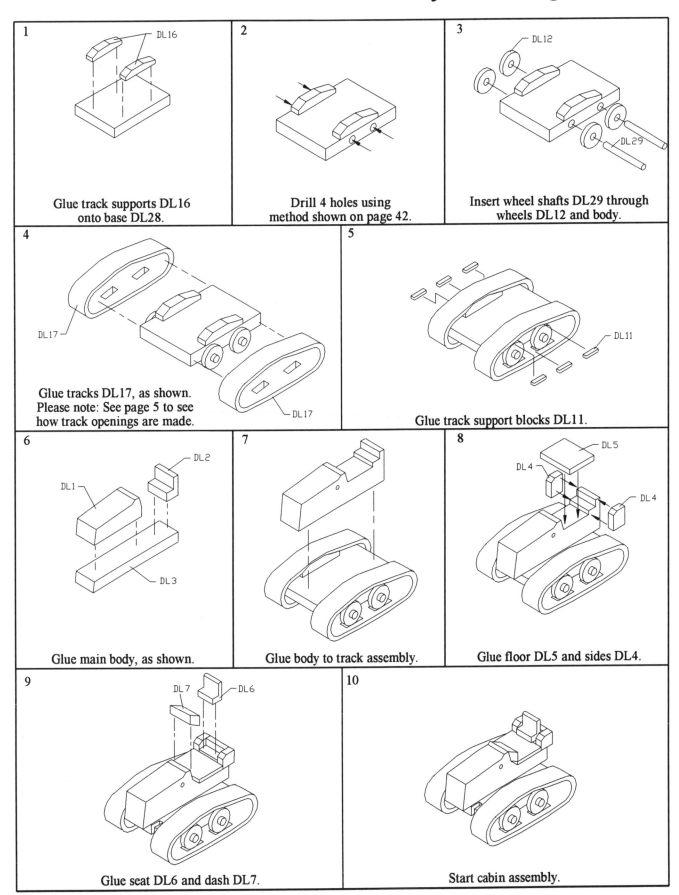

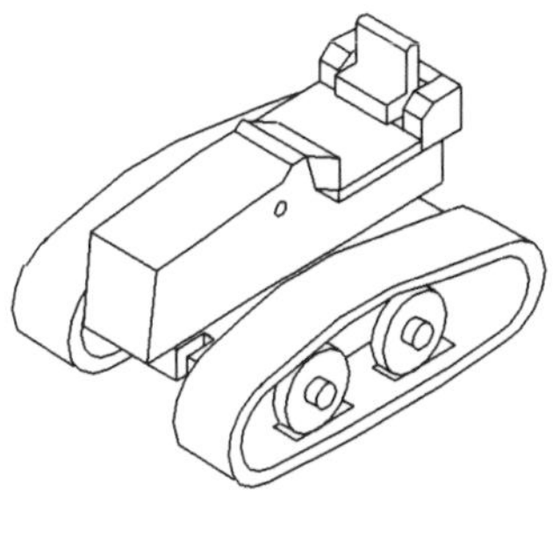

Start cabin assembly.

11

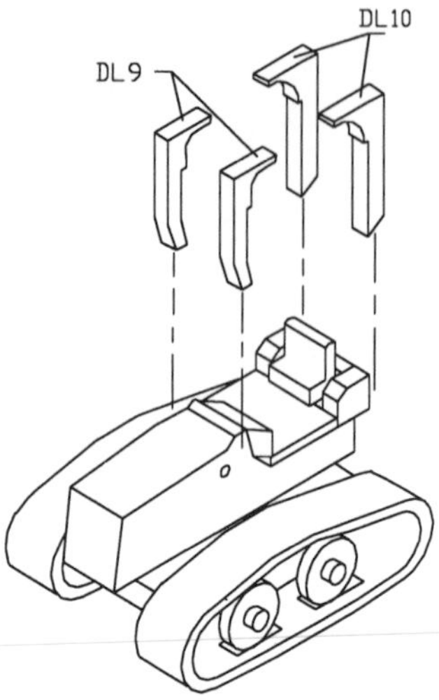

Glue cabin frames DL9 and DL10.

IMPORTANT

Cabin must be completely assembled before glue dries so that you have time to make corrections, if necessary.

12

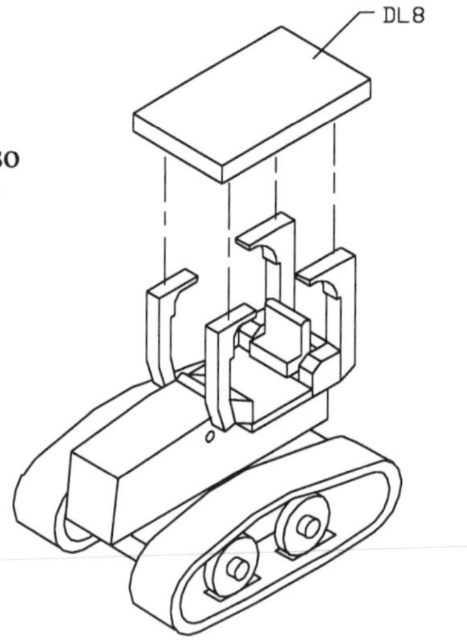

Glue roof assembly DL8.

13

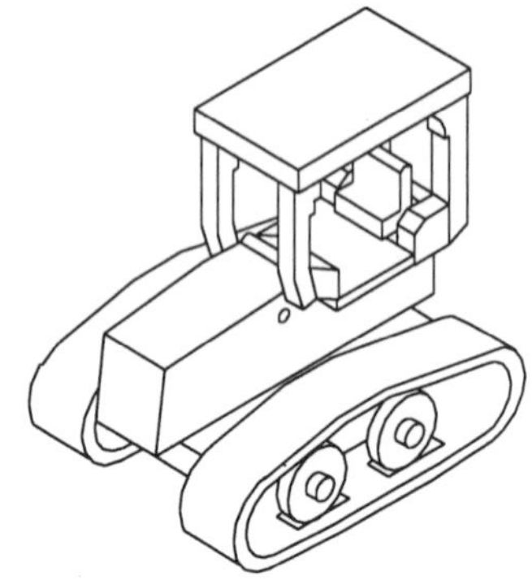

Start shovel assembly.

14

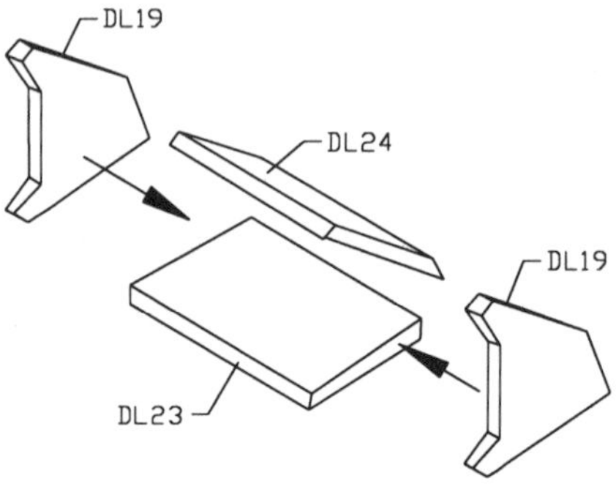

Glue shovel sides DL19 to shovel top DL24 and shovel bottom DL23.

15

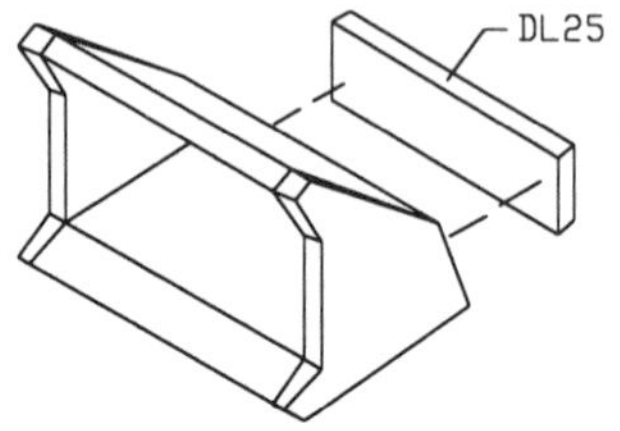

Glue shovel back DL25, as shown. See page 6 for hints on making the shovel assembly.

16

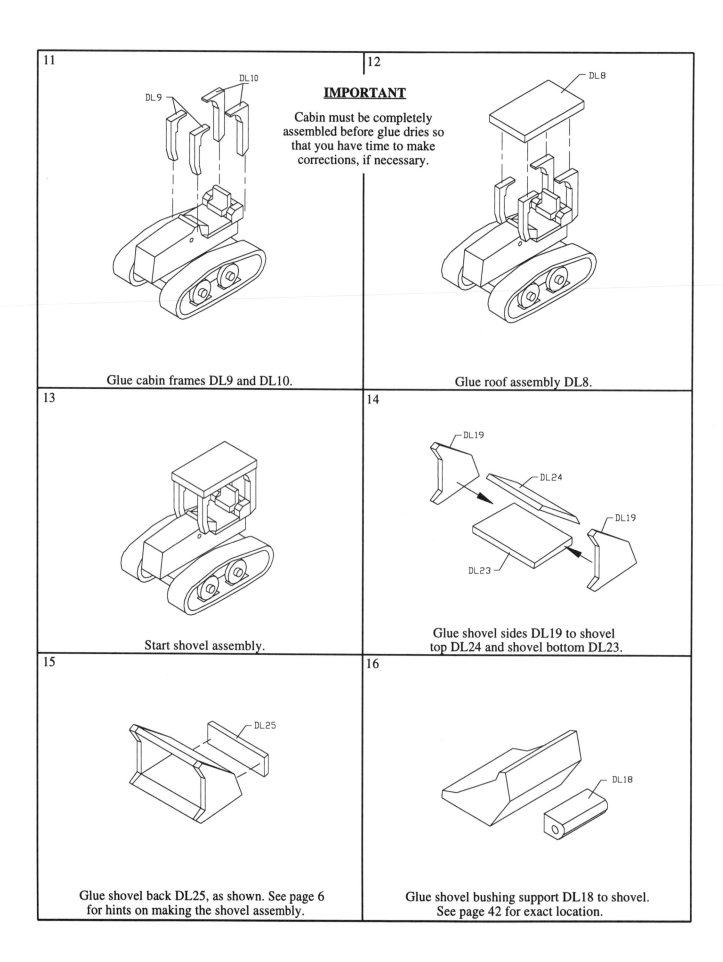

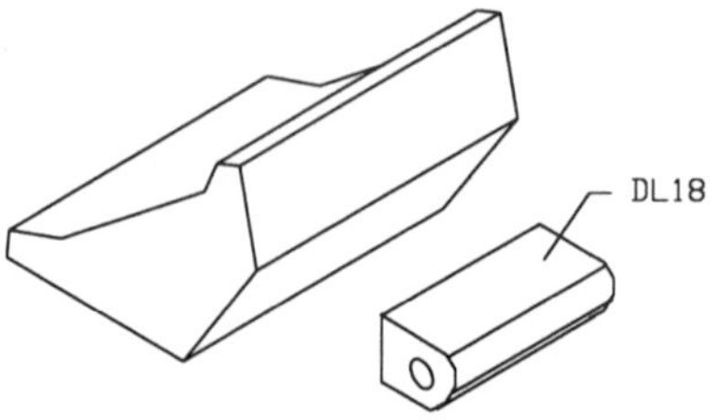

Glue shovel bushing support DL18 to shovel. See page 42 for exact location.

17

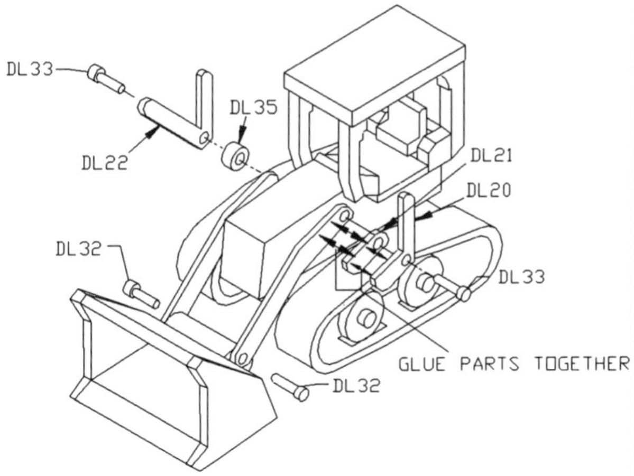

Assemble parts, as shown.

18

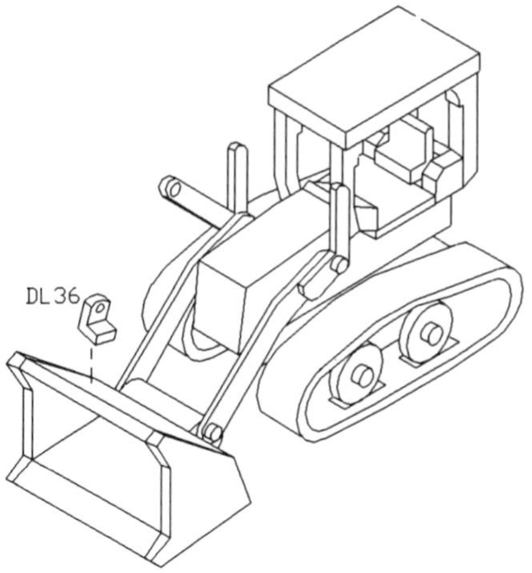

Glue lever block DL36. See page 42 for exact location.

19

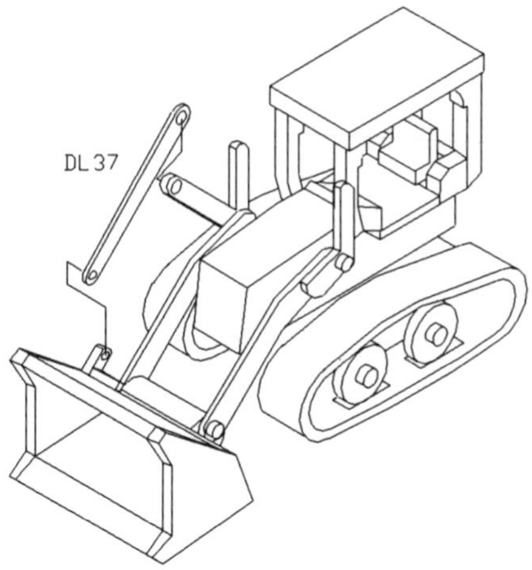

Install lever arm DL37.

20

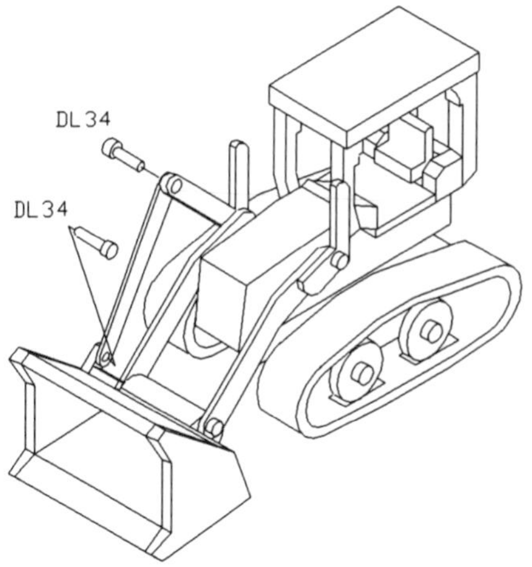

Insert locking pins DL34.

21

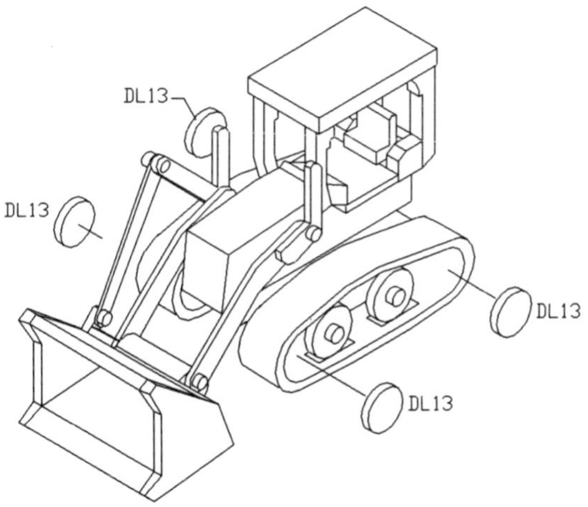

Glue parts, as shown.

22

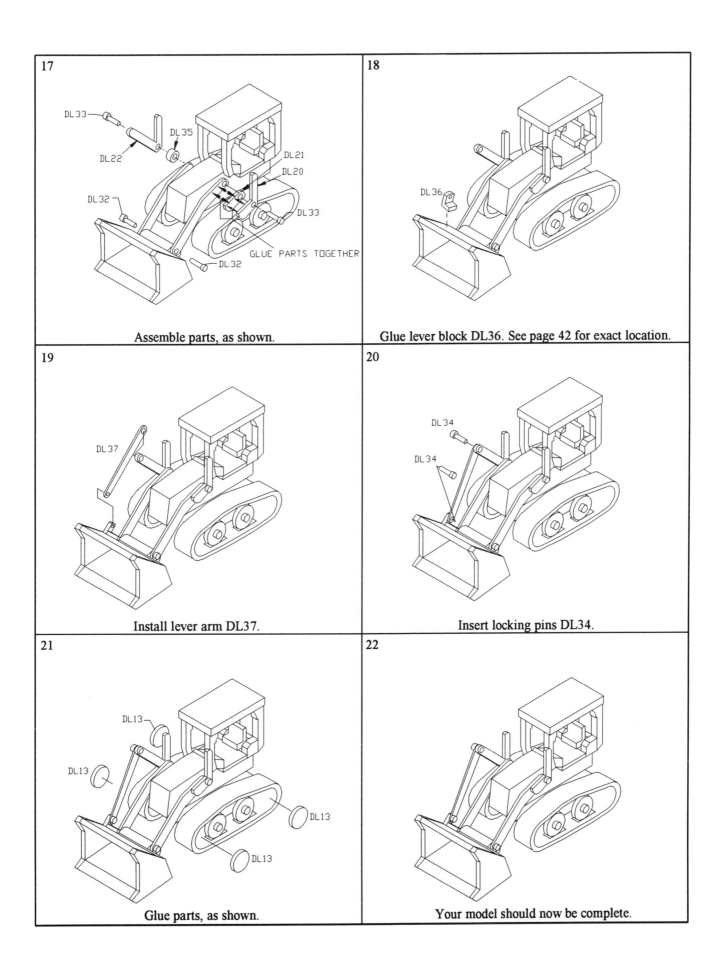

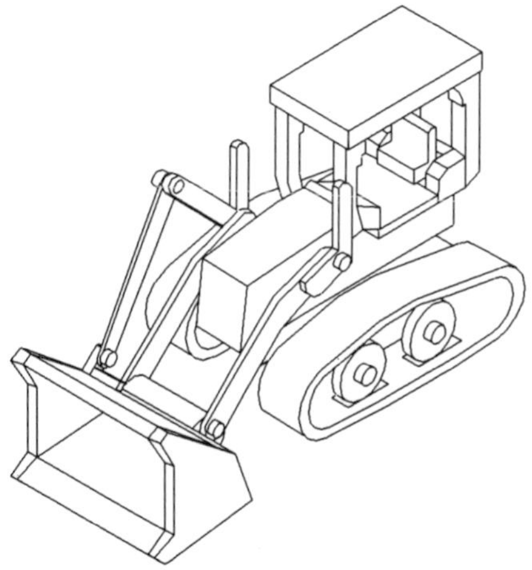

Your model should now be complete.

EXCAVATOR

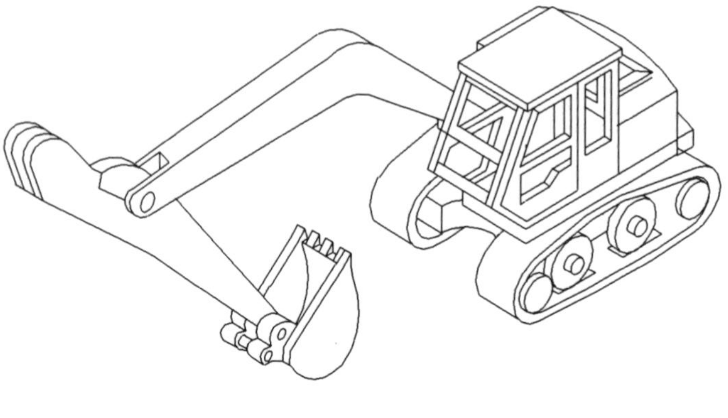

General Instructions - Excavator

1- Start by cutting materials needed by following the list of materials, paying attention to the rough and finished size. **Identify the parts as they are cut.**

Please note: Different types of wood can be used for the various parts. It is suggested, however, that hard wood be used, since many of the parts would be much too fragile if using soft wood. We have used a combination of pine, maple and oak to give the models a nice contrast!

2- Remove the full-size patterns found in the appendix. Cut them out, leaving approximately 1/16" all around, and place on the proper piece of wood. Patterns can be secured to wood using either spray adhesive or rubber ciment. If using the latter, cut and sand the part first to finished size. If drilling is required, mark the hole by inserting a scriber or nail through the pattern into the wood. Remove the pattern before drilling.

You should have no trouble determining which surface to attach most of the patterns. Some parts, however, can be confusing since the pattern could fit on more than one surface. The drawings below indicate exactly which surface to attach the patterns for these parts.

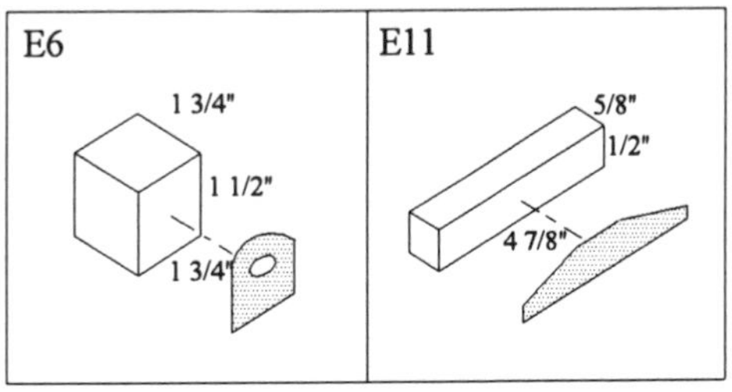

4- Look at the full-size drawing sheets to finish parts E5, E12, and E13.

5- Parts E6 and E19 will need additional cuts and details, please refer to the additional information pages, to complete these parts.

6- Using maple dowels, make all pins, shafts, etc.

7- Follow the assembly drawings to complete your model.

List of materials - Excavator

Part	T	W	L	Material	Qty.	*
E1	3/8"	4"	4"	oak	1	R
E2	3/8"	2"	4"	oak	1	R
E3	3/8"	4"	4"	oak	1	R
E4	3/8"	2 5/8"	3 1/8"	pine	1	F
E5	1/2"	5 9/16"	6 3/8"	pine	1	F
E6	1 1/2"	1 3/4"	1 3/4"	maple	1	R
E7	3/8"	1 7/8"	1 7/8"	maple	4	R
E8	3/8"	3/8"	1 3/4"	oak	2	F
E9	1/4"	1/2"	1 1/4"	pine	6	F
E10	1 1/2"	3 7/8"	5 5/8"	pine	1	R
E11	3/8"	5/8"	4 3/4"	pine	2	F
E12	1/4"	1 7/8"	3 7/16"	pine	1	F
E13	1 1/4"	3 7/8"	7"	pine	1	F
E14	3/8"	2 5/8"	2 5/8"	maple	1	R
E15	1/2"	2 3/8"	2 3/8"	maple	1	R
E17	1"	3 3/4"	11"	maple	1	R
E18	1/4"	2"	3 1/8"	maple	2	R
E19	1 3/4"	2 1/2"	3 1/2"	maple	1	R
E20	1/2"	1 1/2"	5 5/8"	oak	1	R
E21	1 1/2"	2 1/2"	8 3/4"	pine	2	R
E22	3/8"	1 3/8"	7"	maple	2	R
E23	3/8"	2"	7 7/8"	maple	1	R
E25	1/4"	1 9/16" DIA.		maple	4	R

R = Rough size
F = Finished size

T = Thickness
W = Width
L = Length

Instructions:

R= Rough sizes, the material is cut oversized so you have ample room to apply the pattern on the surface. Sanding is not required at this point.

F = Finished Size: Cut and sand parts to finished size.

Full-Sized Patterns: Set One

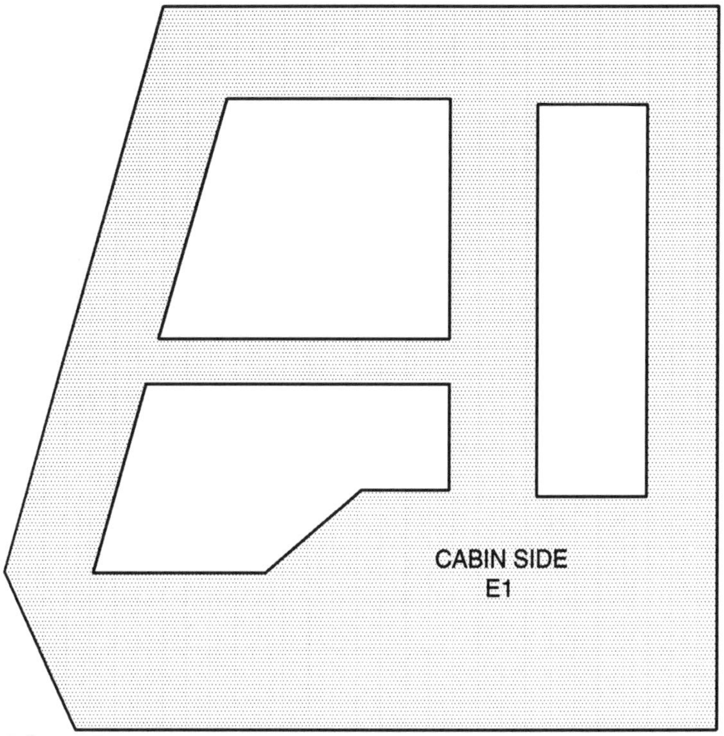

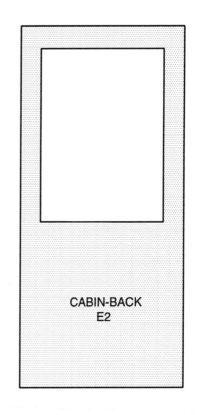

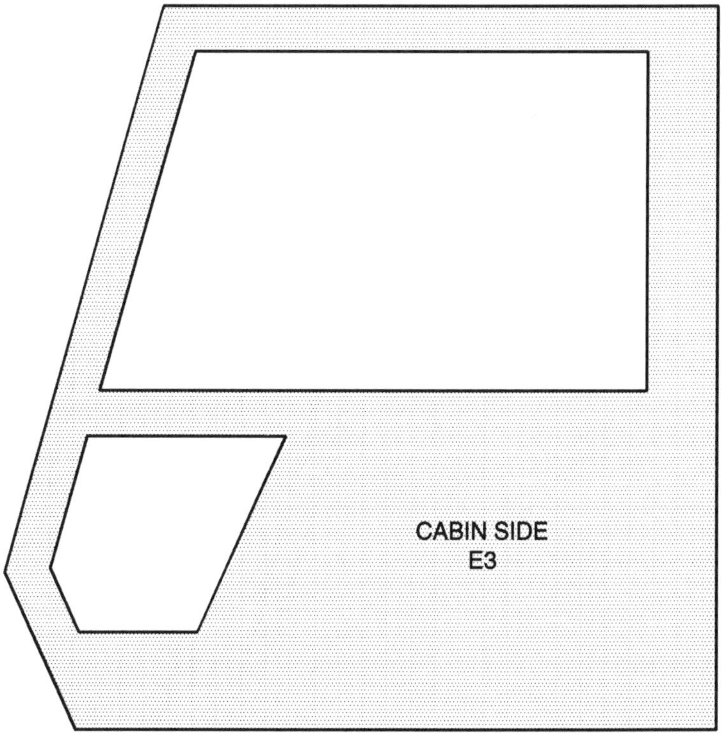

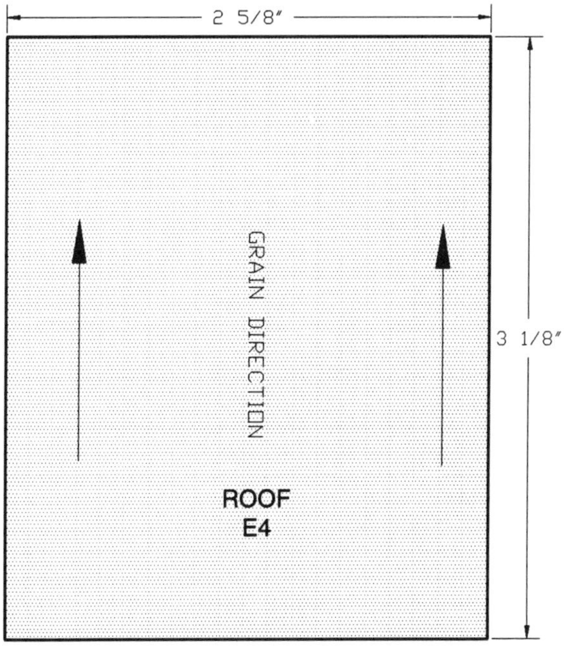

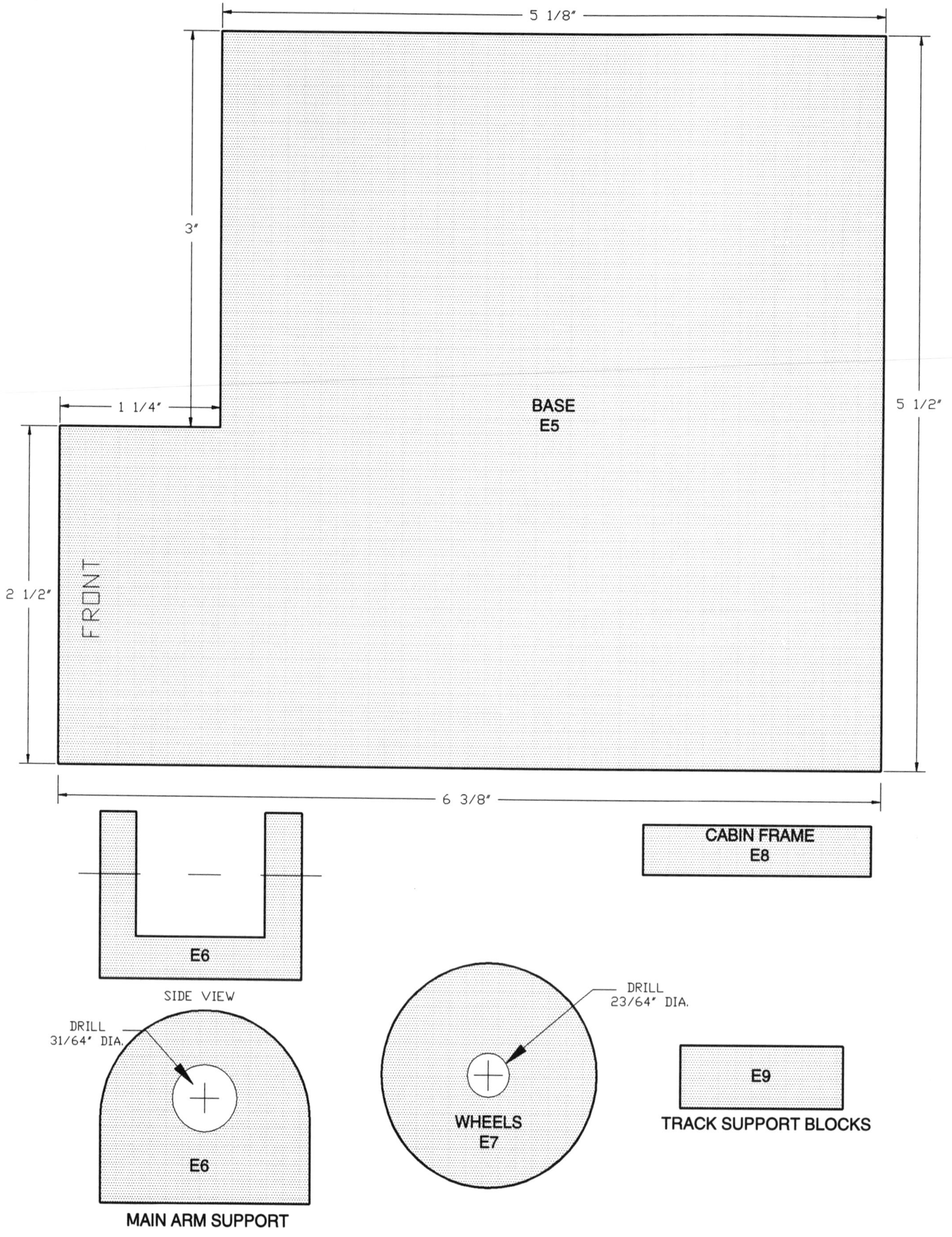
5 1/8"
3"
1 1/4"
BASE
E5
5 1/2"
2 1/2"
FRONT
6 3/8"
CABIN FRAME
E8
E6
SIDE VIEW
DRILL
31/64" DIA.
E6
MAIN ARM SUPPORT
DRILL
23/64" DIA.
WHEELS
E7
E9
TRACK SUPPORT BLOCKS

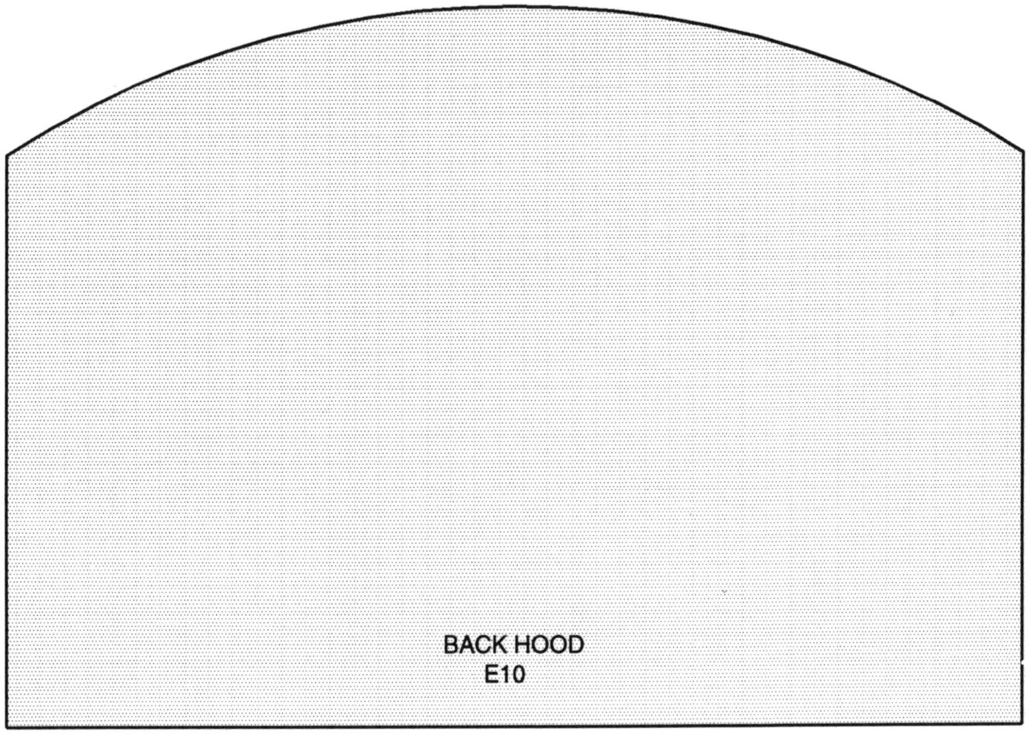
BACK HOOD
E10

TRACK SUPPORT
E11
SIDE VIEW

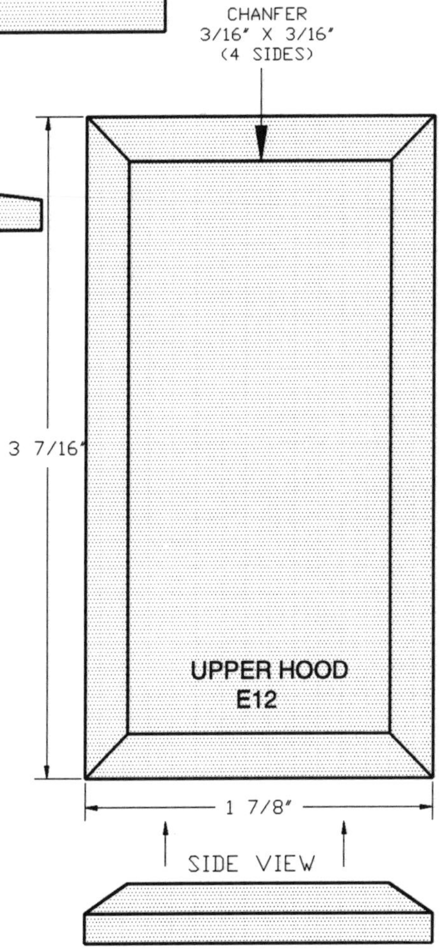
CHANFER
3/16" X 3/16"
(4 SIDES)
3 7/16"
UPPER HOOD
E12
1 7/8"
SIDE VIEW

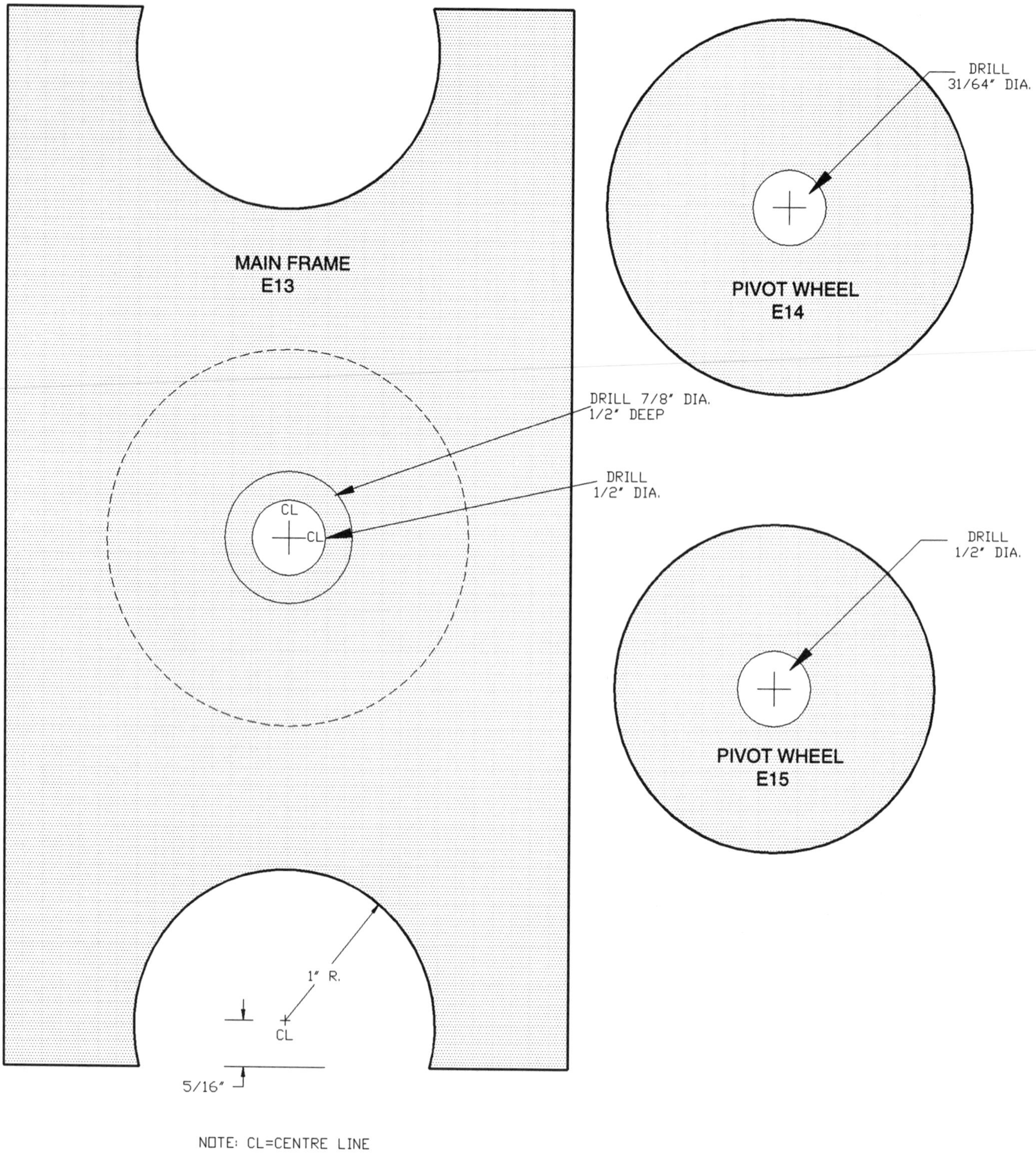
MAIN FRAME
E13
DRILL 7/8" DIA.
1/2" DEEP
DRILL
1/2" DIA.
CL
CL
1" R.
CL
5/16"
NOTE: CL=CENTRE LINE
DRILL
31/64" DIA.
PIVOT WHEEL
E14
DRILL
1/2" DIA.
PIVOT WHEEL
E15

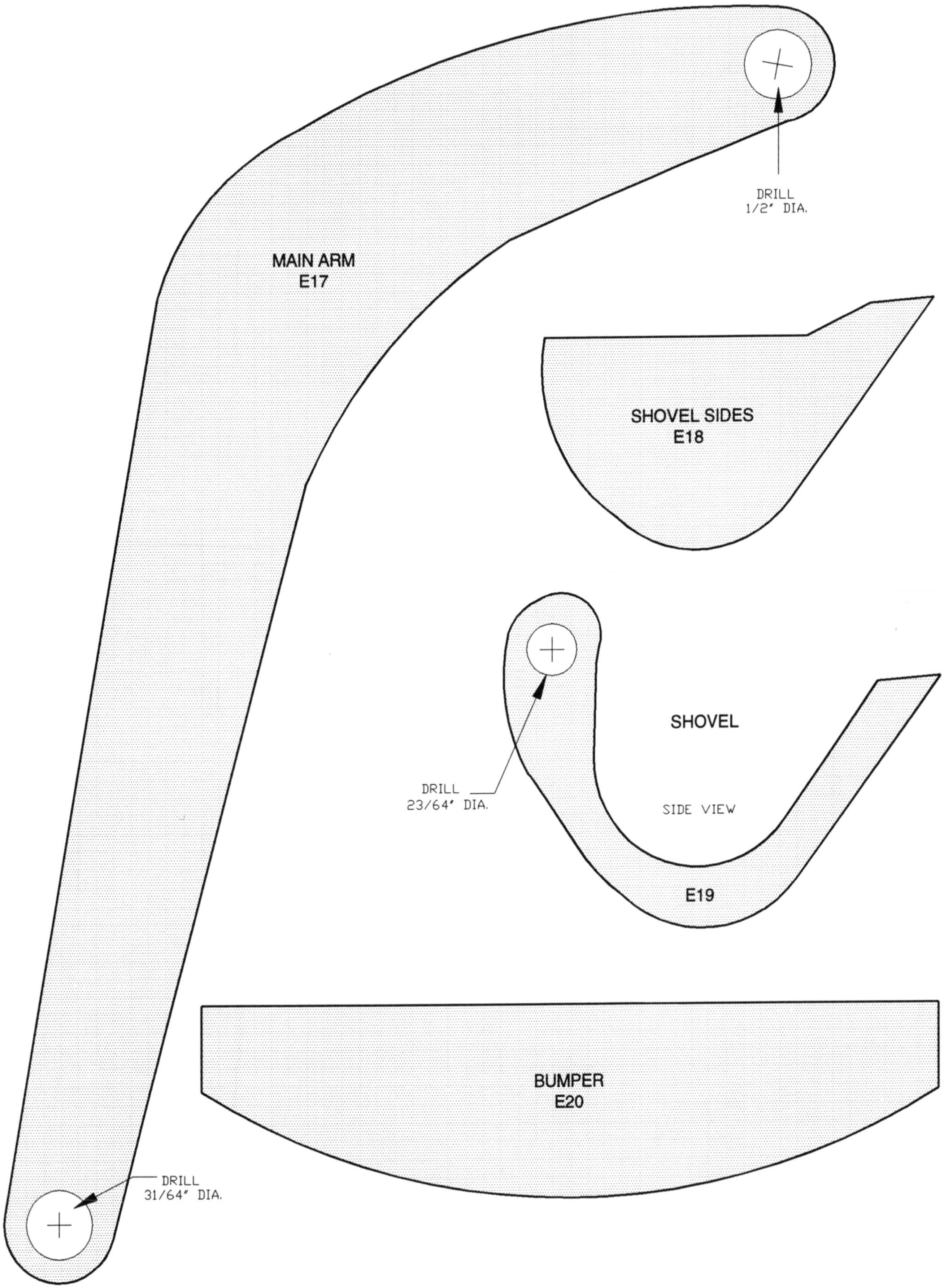
DRILL
1/2" DIA.
MAIN ARM
E17
SHOVEL SIDES
E18
SHOVEL
DRILL
23/64" DIA.
SIDE VIEW
E19
BUMPER
E20
DRILL
31/64" DIA.

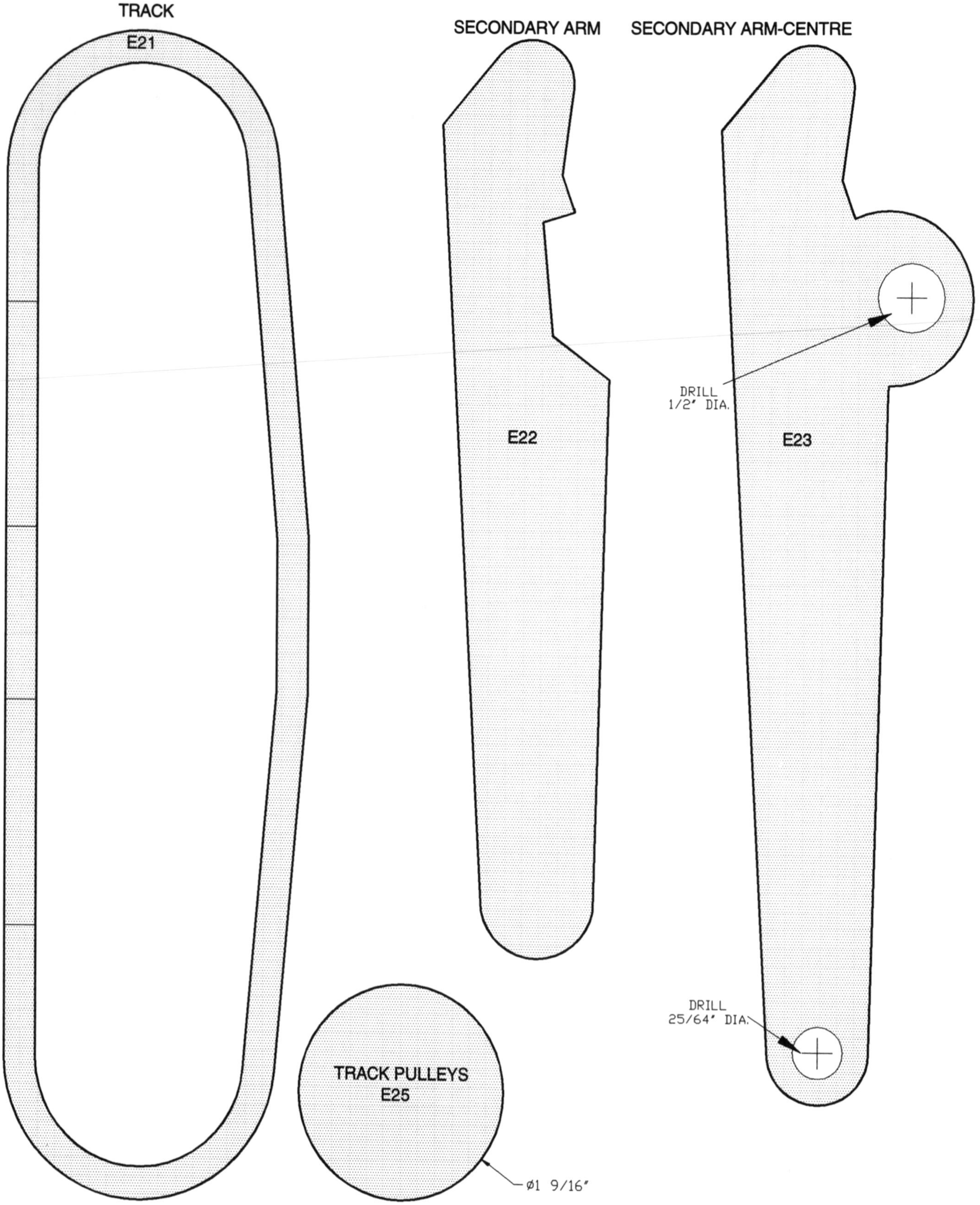
TRACK
E21
SECONDARY ARM
E22
SECONDARY ARM-CENTRE
DRILL
1/2" DIA.
E23
DRILL
25/64" DIA.
TRACK PULLEYS
E25
Ø1 9/16"

PIN
(1 REQUIRED)
E26
1/2" DIA. DOWEL
1"
PIN
(1 REQUIRED)
E27
3/8" DIA. DOWEL
1 3/4"
WHEEL SHAFTS
(2 REQUIRED)
E28
3/8" DIA. DOWEL
4 7/8"
SIDE VIEW
MAIN PIN
(1 REQUIRED)
E29
3/4" DIA. DOWEL
1/2" DIA. DOWEL
DRILL 1/2" DIA.
3/8" DEEP
GLUE IN
SIDE VIEW
1/2"
2 1/2"
SIDE VIEW
ASSEMBLED PART
PIN
(1 REQUIRED)
E31
1/2" DIA. DOWEL
1 3/4"

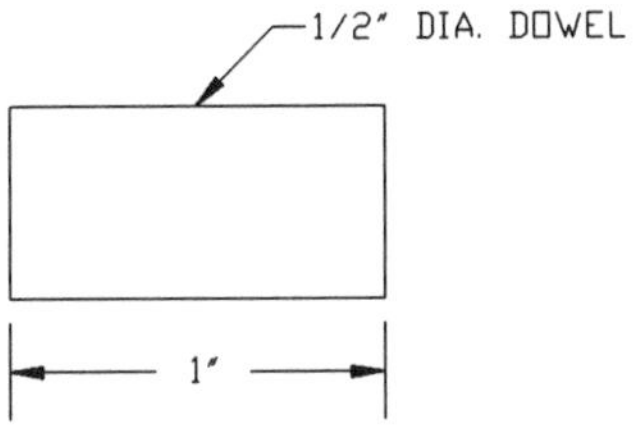

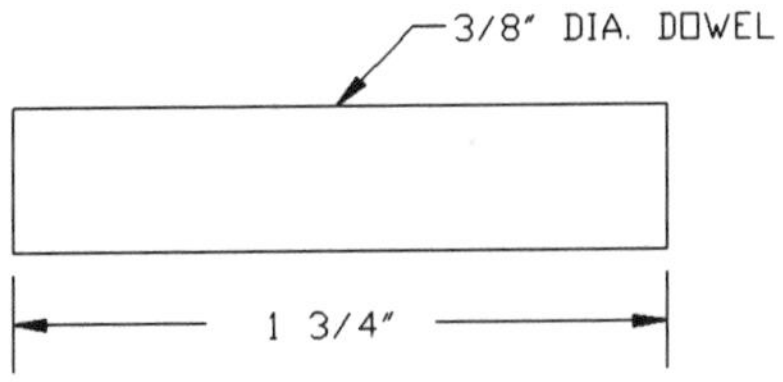

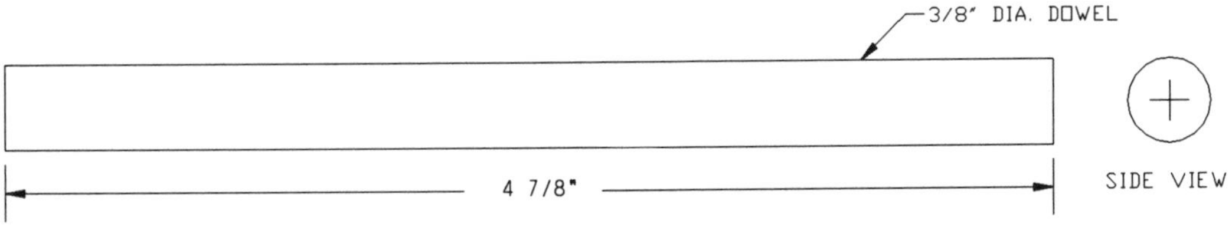

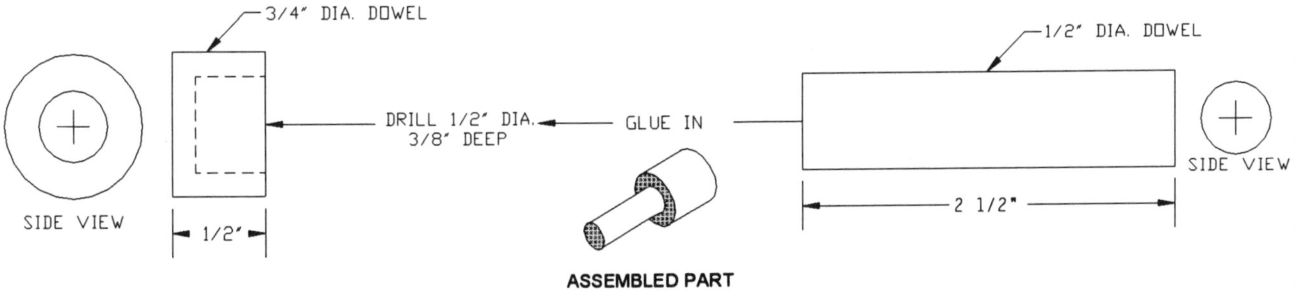

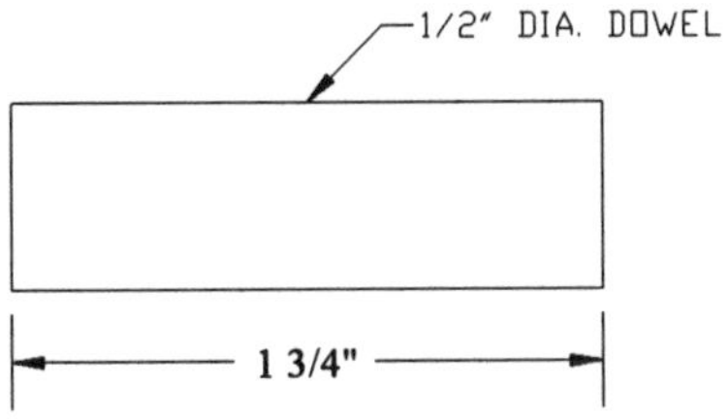

Additional Information - Excavator

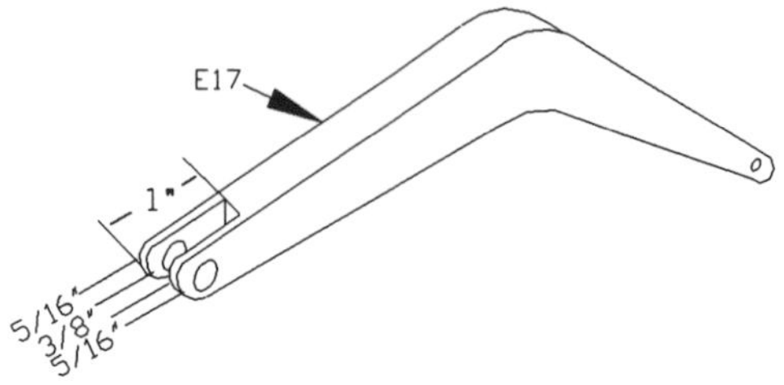

Using your Scroll Saw, cut groove in part E17, as shown.

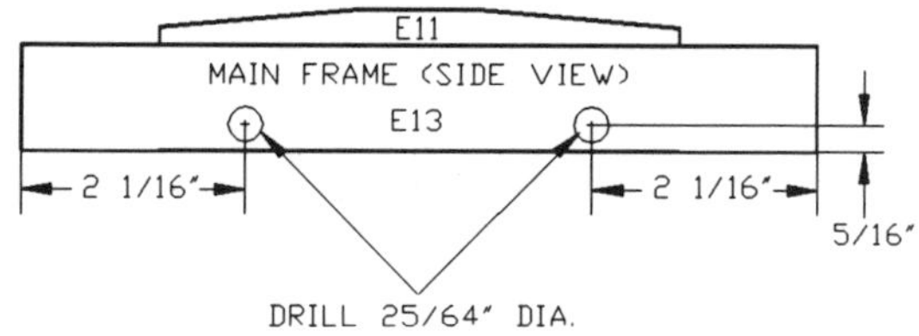

Drill holes into main frame E13, as shown.

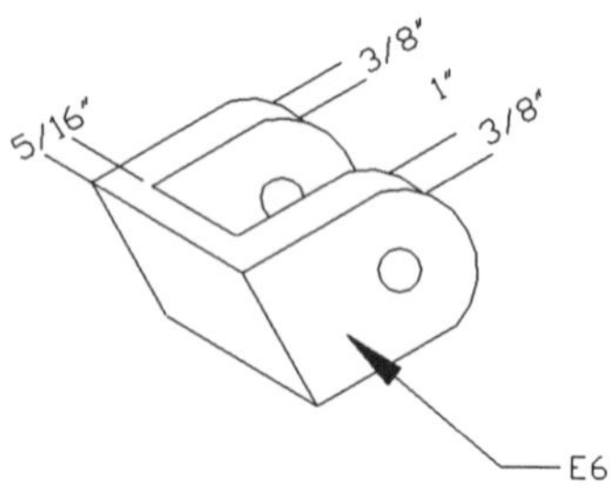

Using your Scroll Saw, cut groove in part E6, as shown.

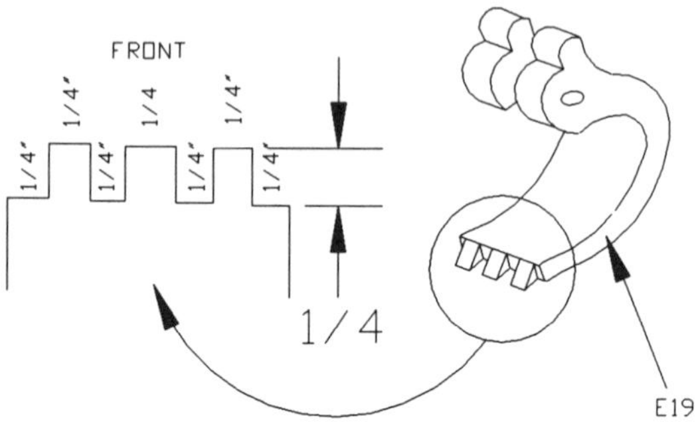

Using your Scroll Saw, cut grooves in part E19, as shown.

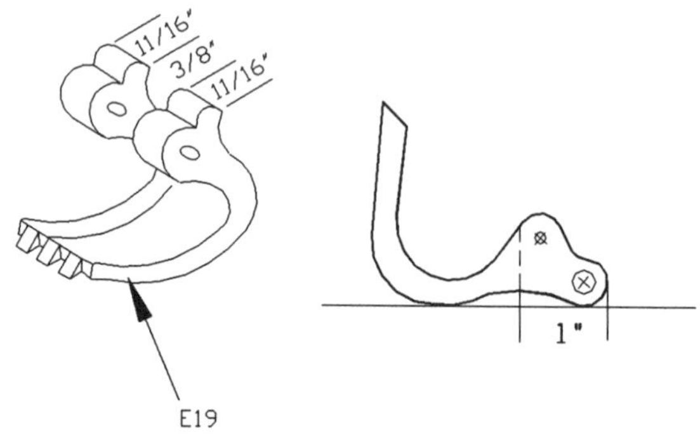

Using your Scroll Saw, cut groove in part E19, as shown.

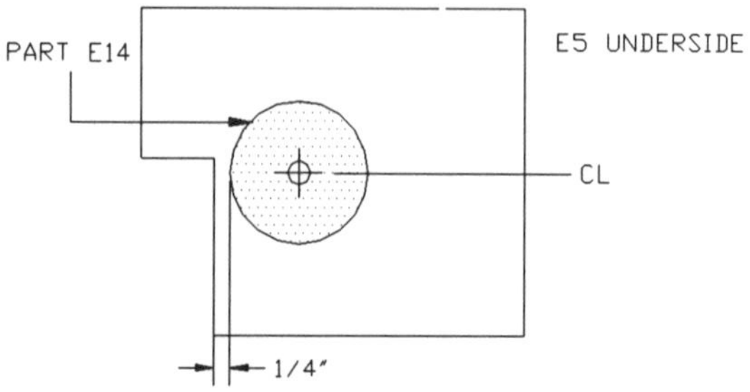

Drawing showing where to glue pivot wheel E14 onto base E5.

Excavator - Assembly Drawings

1

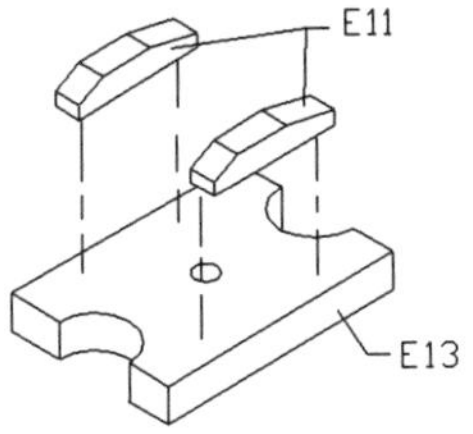

Glue track supports E11 onto base E13.

2

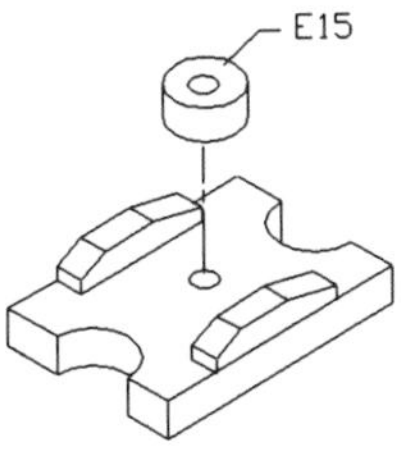

Glue pivot wheel E15 making sure holes are in line.

3

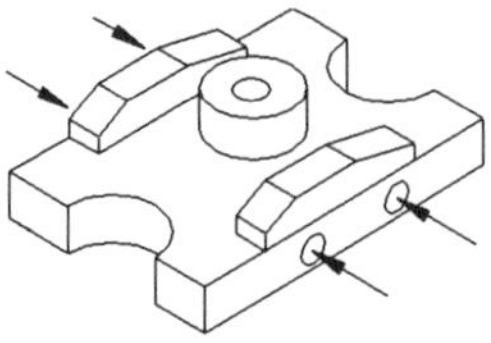

Drill 4 holes using method shown on page 56.

4

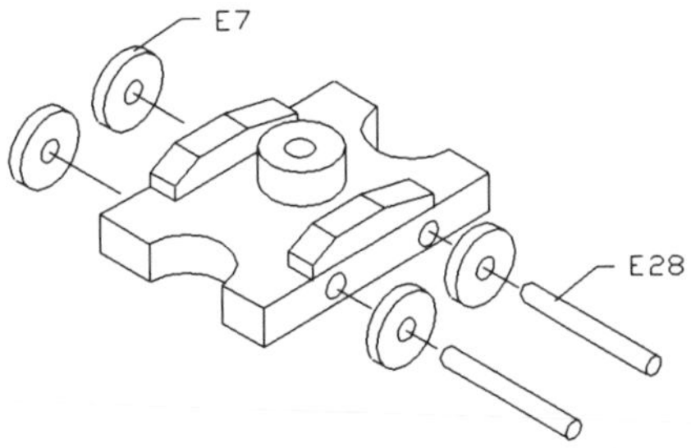

Insert wheel shaft E28 through wheels E7 and body.

5

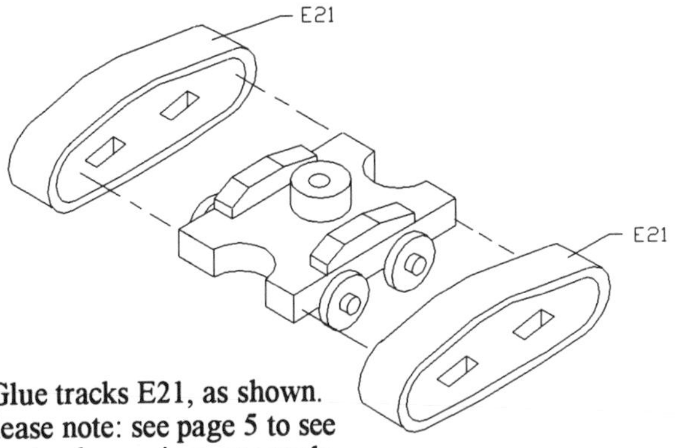

Glue tracks E21, as shown. Please note: see page 5 to see how track openings are made.

6

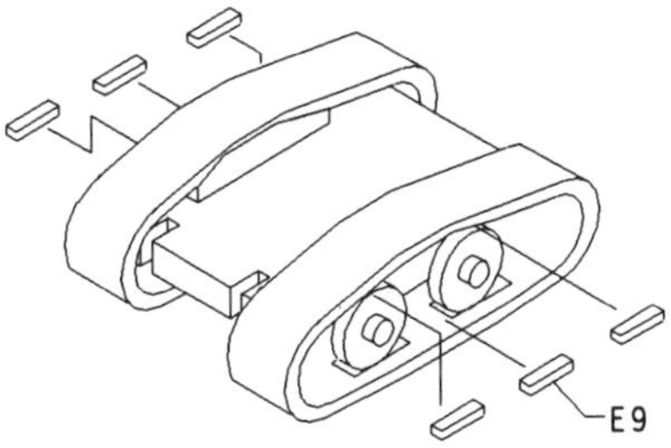

Glue track support blocks E9.

7

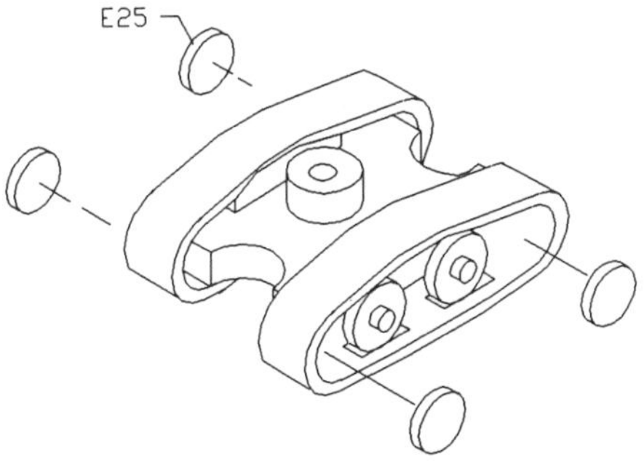

Glue track pulleys E25.

8

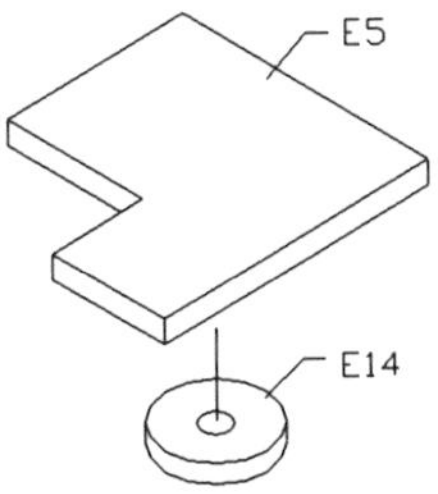

Glue pivot wheel E14 to base E5. See page 56 for exact location.

9

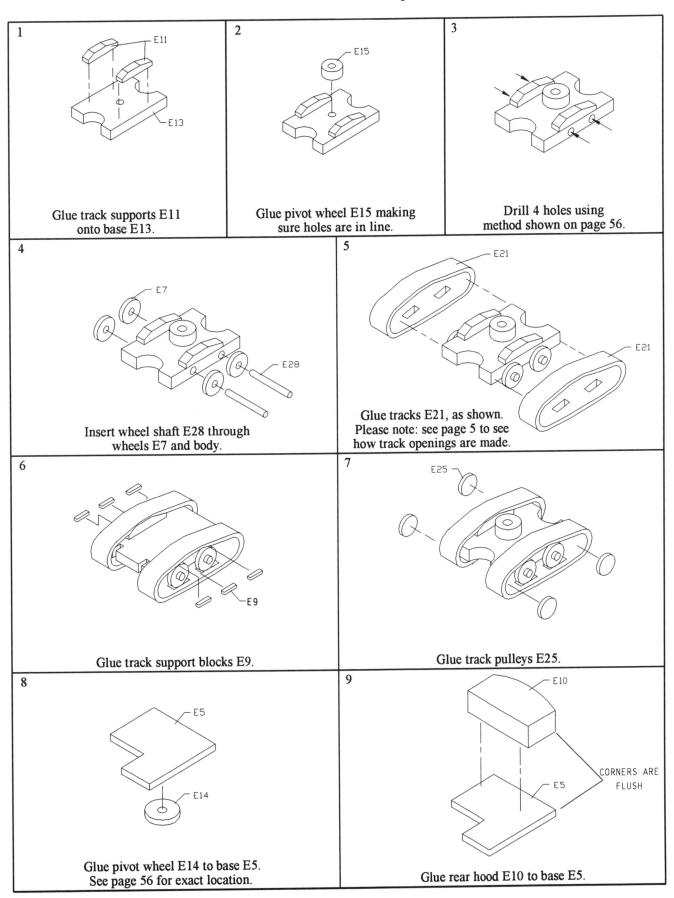

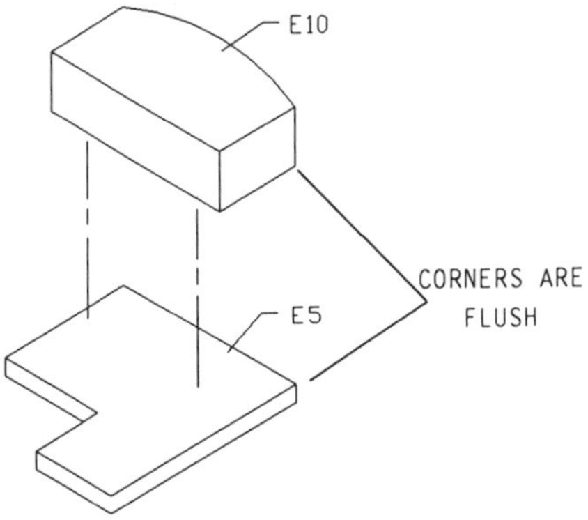

Glue rear hood E10 to base E5.

10

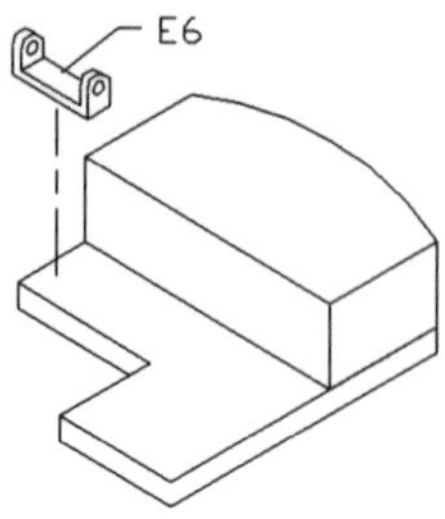

Glue main arm support E6.

11

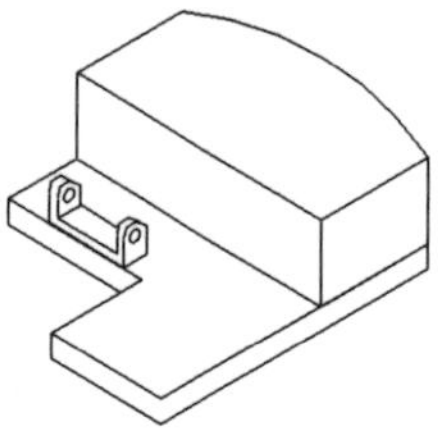

Use this drawing as a guide for part E6.

12

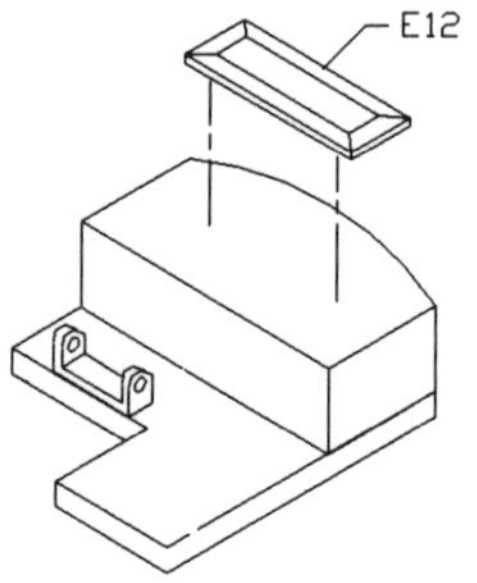

Glue upper hood E12.

13

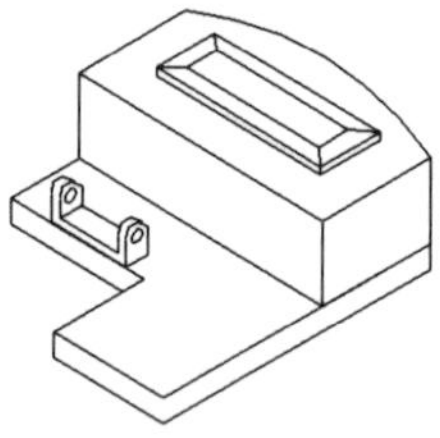

Start working on arm assembly.

14

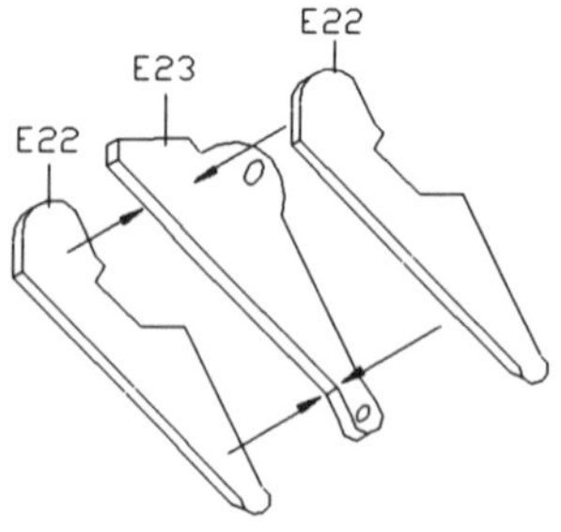

Glue secondary arm assembly, as shown.

15

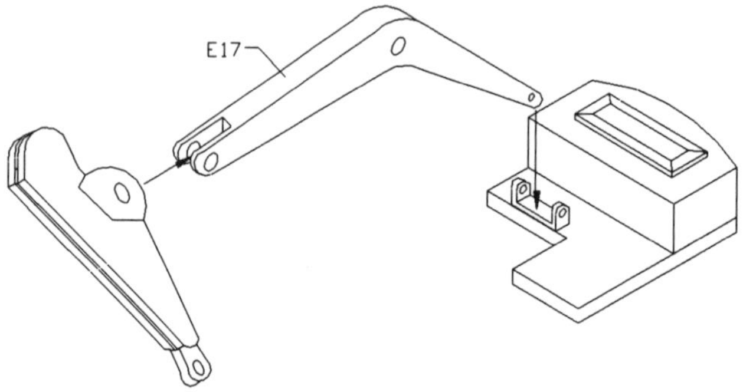

Slide secondary arm into main arm E17.

16

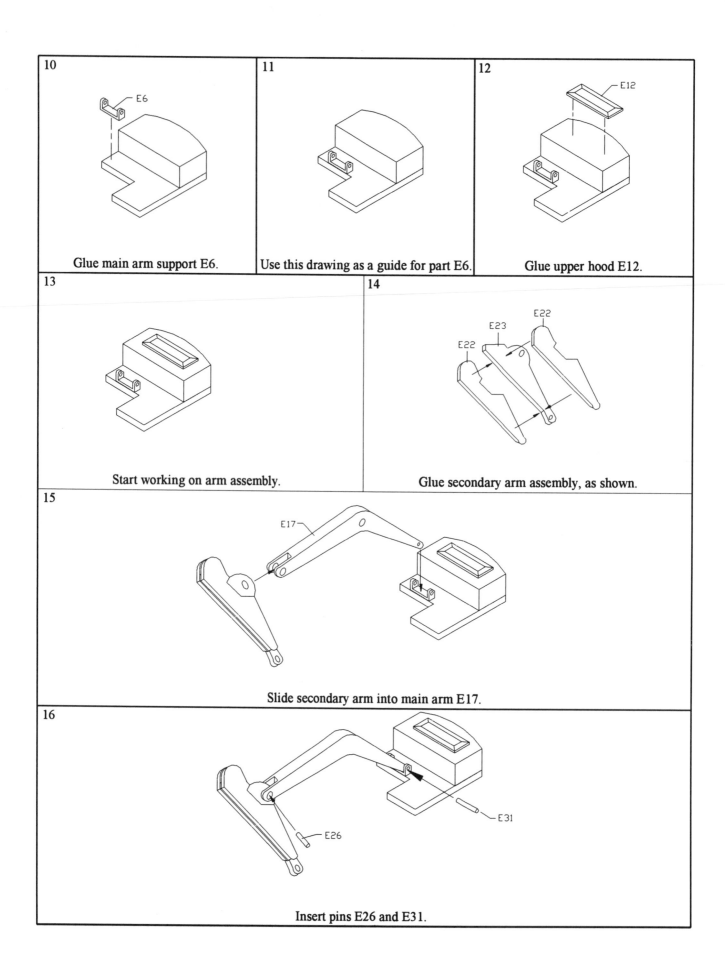

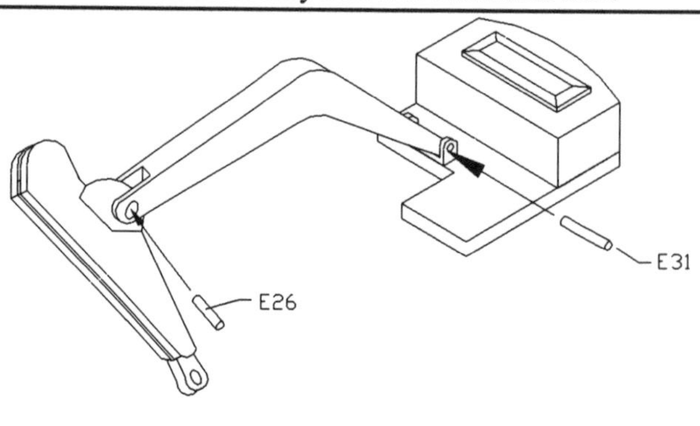

Insert pins E26 and E31.

17

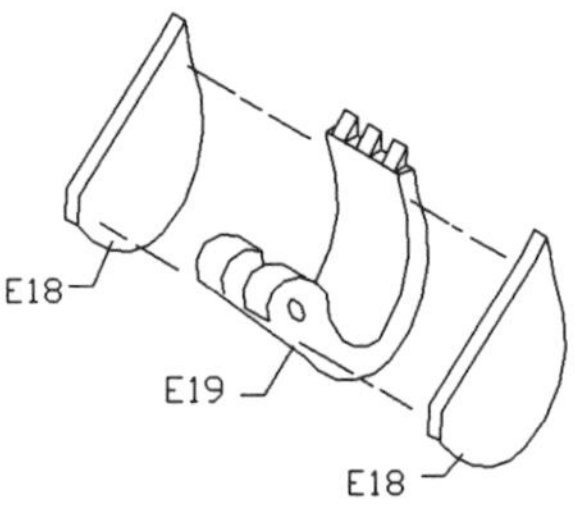

Glue shovel sides E18 onto shovel E19.

18

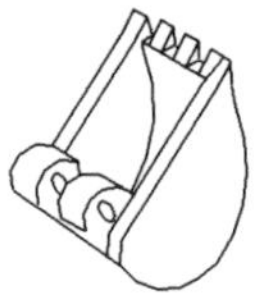

See page 56 for details on shovel assembly.

19

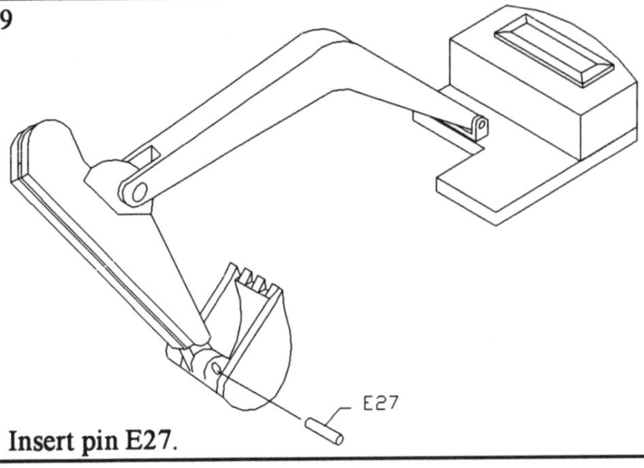

Insert pin E27.

20

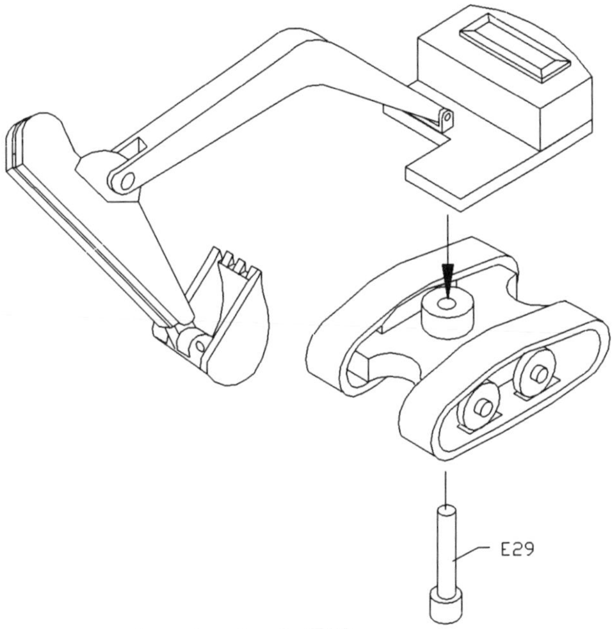

Insert main pin E29.

21

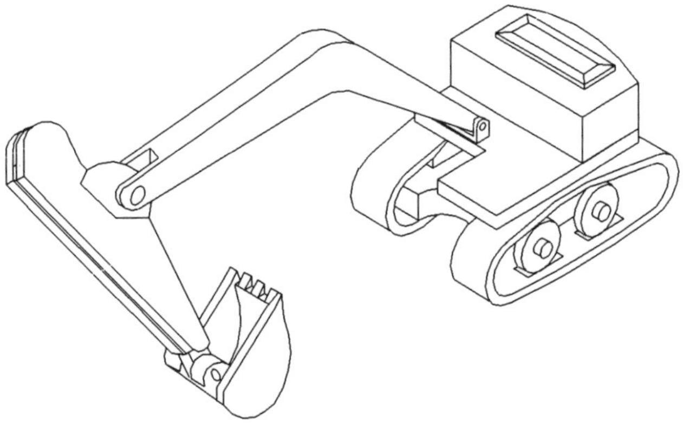

Start cabin assembly.

22

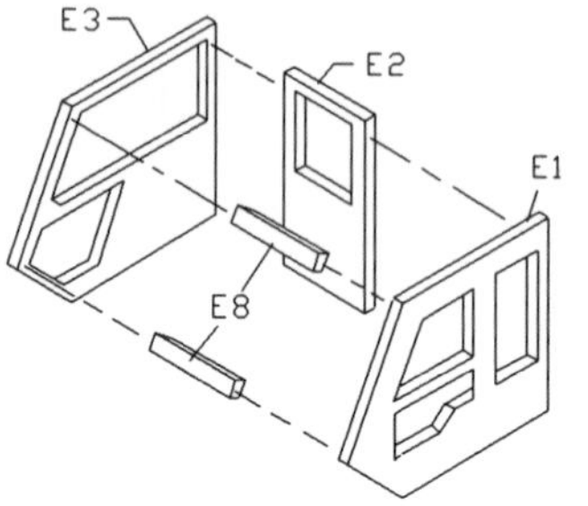

Assemble cabin, as shown.

23

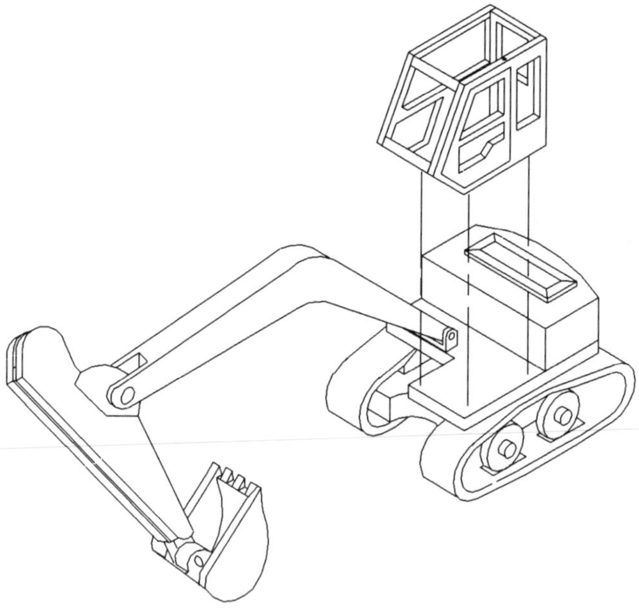

Glue cabin onto base.

24

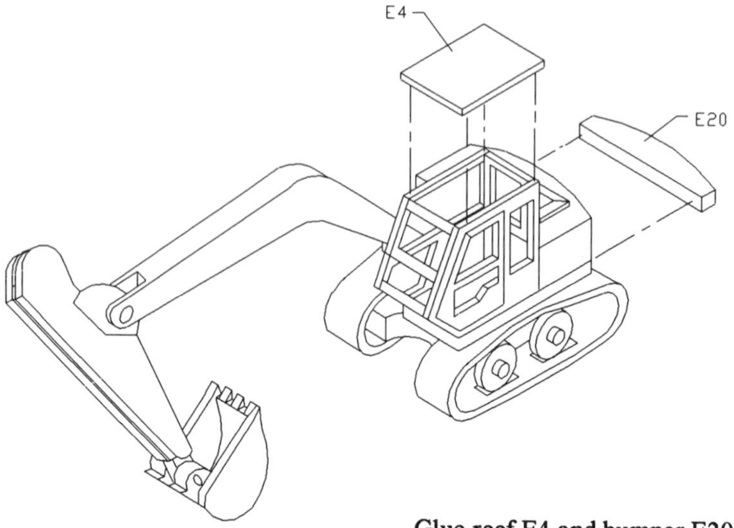

Glue roof E4 and bumper E20.

25

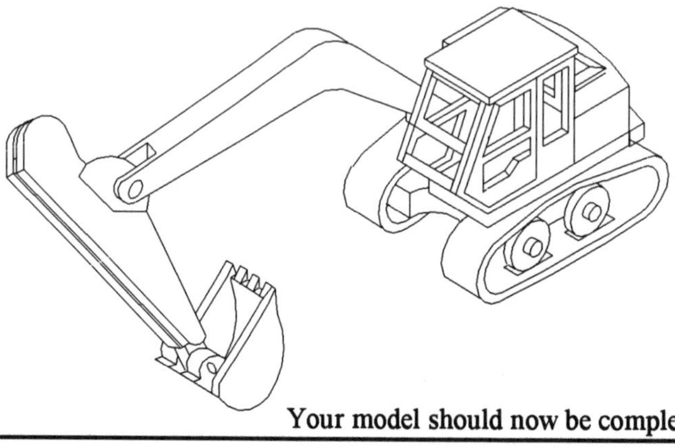

Your model should now be complete.

GRADER

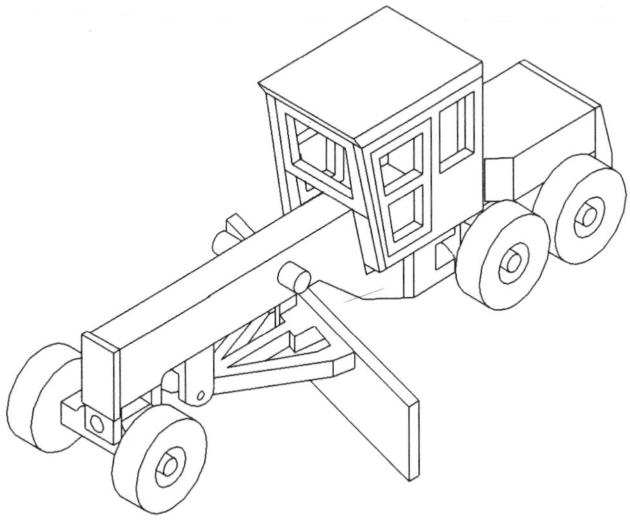

General Instructions - Grader

1- Start by cutting materials needed by following the list of materials, paying attention to the rough and finished size. **Identify the parts as they are cut.**

Please note: Different types of wood can be used for the various parts. It is suggested, however, that hard wood be used, since many of the parts would be much too fragile if using soft wood. We have used a combination of pine, maple and oak to give the models a nice contrast!

2- Remove the full-size patterns found in the appendix. Cut them out, leaving approximately 1/16" all around, and place on the proper piece of wood. Patterns can be secured to wood using either spray adhesive or rubber ciment. If using the latter, cut and sand the part first to finished size. If drilling is required, mark the hole by inserting a scriber or nail through the pattern into the wood. Remove the pattern before drilling.

You should have no trouble determining which surface to attach most of the patterns. Some parts, however, can be confusing since the pattern could fit on more than one surface. The drawings below indicate exactly which surface to attach the patterns for these parts.

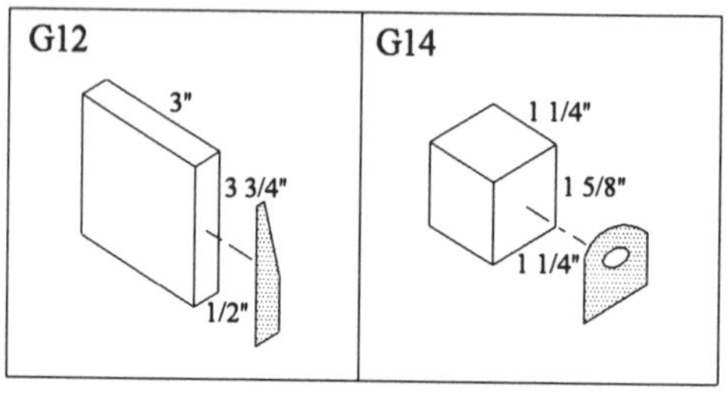

3- Look at the full-size drawing sheets to finish parts G3, G8, G9, G10 and G13.

4- Parts G6, G14, G15 and G16 will need additional cuts and details, please refer to the additional information pages, to complete these parts.

5- Using maple dowels, make all pins, shafts, etc.

6- Follow the assembly drawings to complete your model.

List of Materials - Grader

Part	T	W	L	Material	Qty.	*
G1	3/8"	3 5/8"	3 7/8"	oak	2	R
G2	3/8"	2 3/8"	3 7/8"	oak	1	R
G3	3/8"	3 1/4"	3 7/8"	pine	1	F
G4	3/8"	2 3/8"	3"	oak	1	R
G5	3/8"	3"	2 13/16"	pine	1	F
G6	3/4"	3 1/8"	5 3/8"	pine	1	R
G7	1/4"	1 1/8"	4 7/8"	maple	1	R
G8	1 1/2"	3"	4 3/4"	pine	1	F

Part	T	W	L	Material	Qty.	*
G9	1 1/8"	3"	5 3/4"	pine	1	F
G10	1/2"	3"	5 3/4"	pine	2	F
G11	1/4"	2"	6"	maple	1	F
G12	1/2"	3"	3 3/4"	pine	1	F
G13	1 1/4"	3 1/8" DIA.		oak	4	F
G14	1 1/4"	1 1/4"	1 5/8"	maple	1	R
G15	1"	3 1/4"	2 3/8"	pine	1	R
G16	1 1/4"	4"	12 1/2"	pine	1	R

R = Rough size
F = Finished size

T = Thickness
W = Width
L = Length

Instructions:

R= Rough sizes, the material is cut oversized so you have ample room to apply the pattern on the surface. Sanding is not required at this point.

F = Finished Size: Cut and sand parts to finished size.

Full-Sized Patterns: Set One

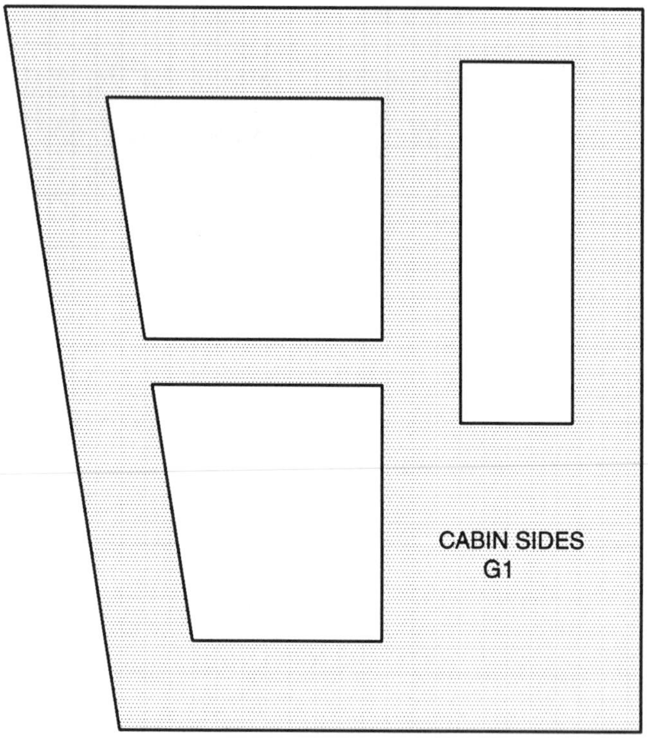

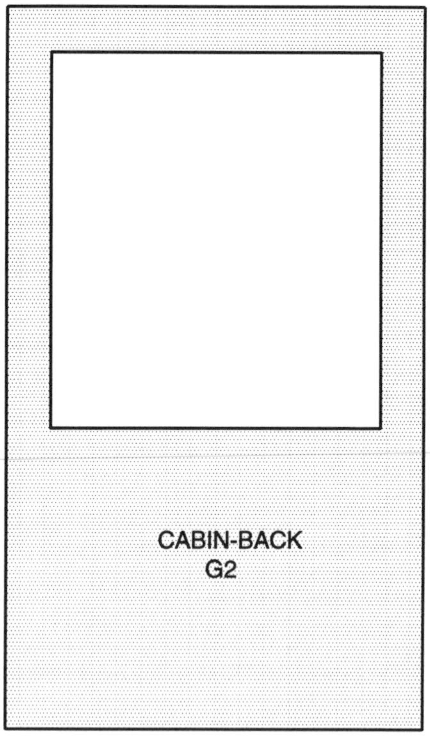

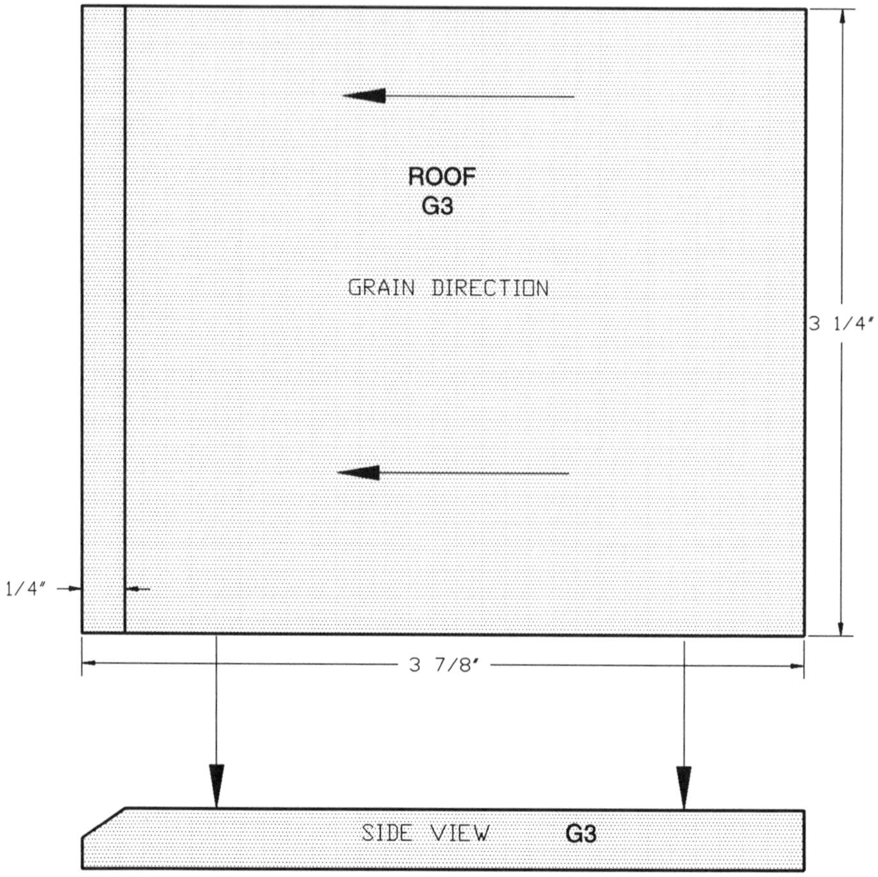

2 13/16"
3"
FLOOR
G5
FRONT
GRAIN DIRECTION

SHOVEL FRAME ASSEMBLY
CONTROL ARM LEVER
DRILL
3/8" DIA.
SIDE VIEW
G7
G6
G6
DRILL
25/64" DIA.
DRILL
23/64" DIA.
1 1/2" DEEP

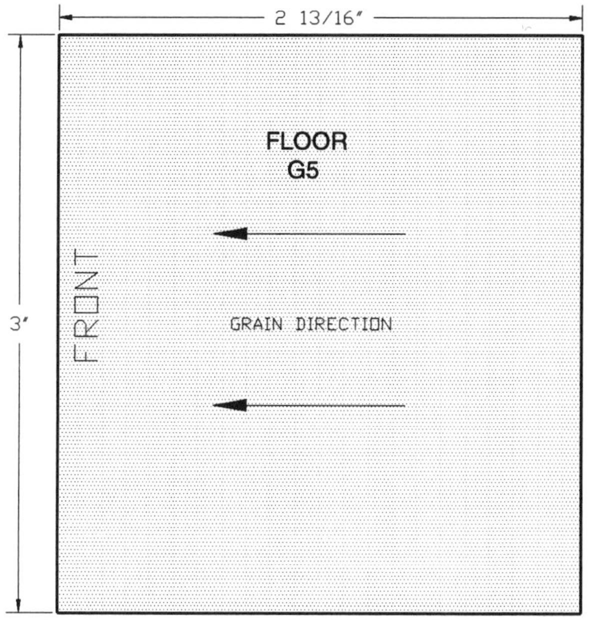

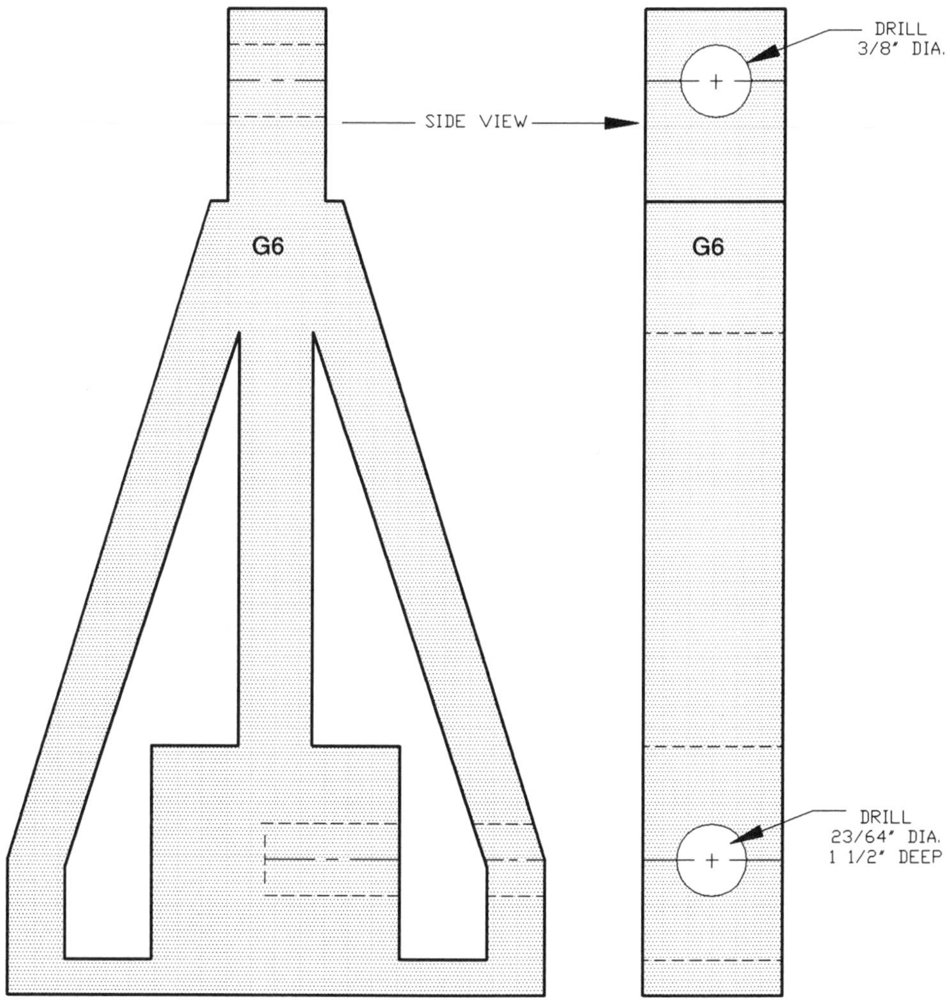

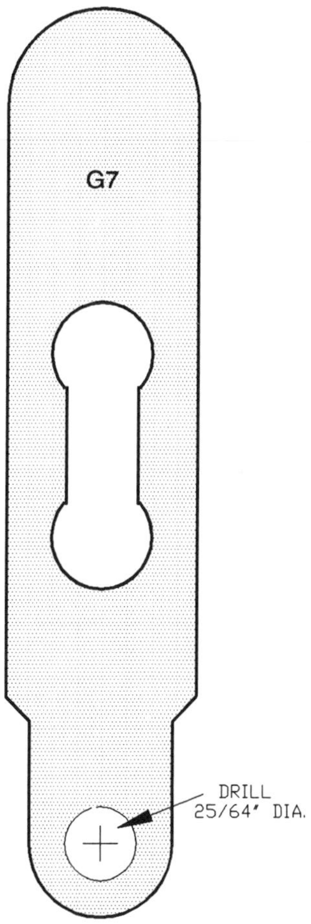

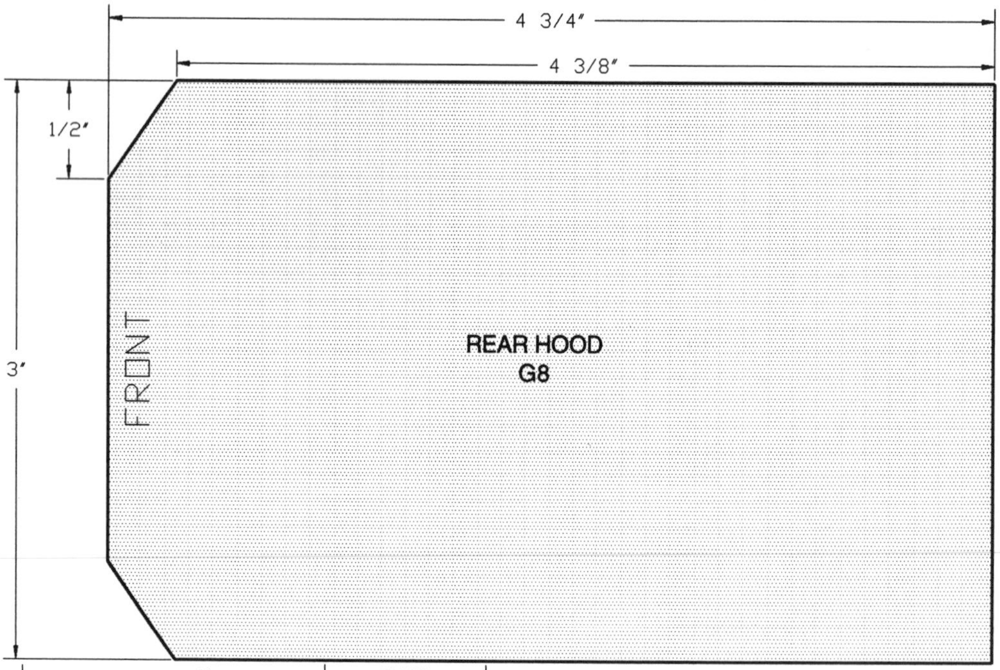
4 3/4″
4 3/8″
1/2″
3″
FRONT
REAR HOOD
G8

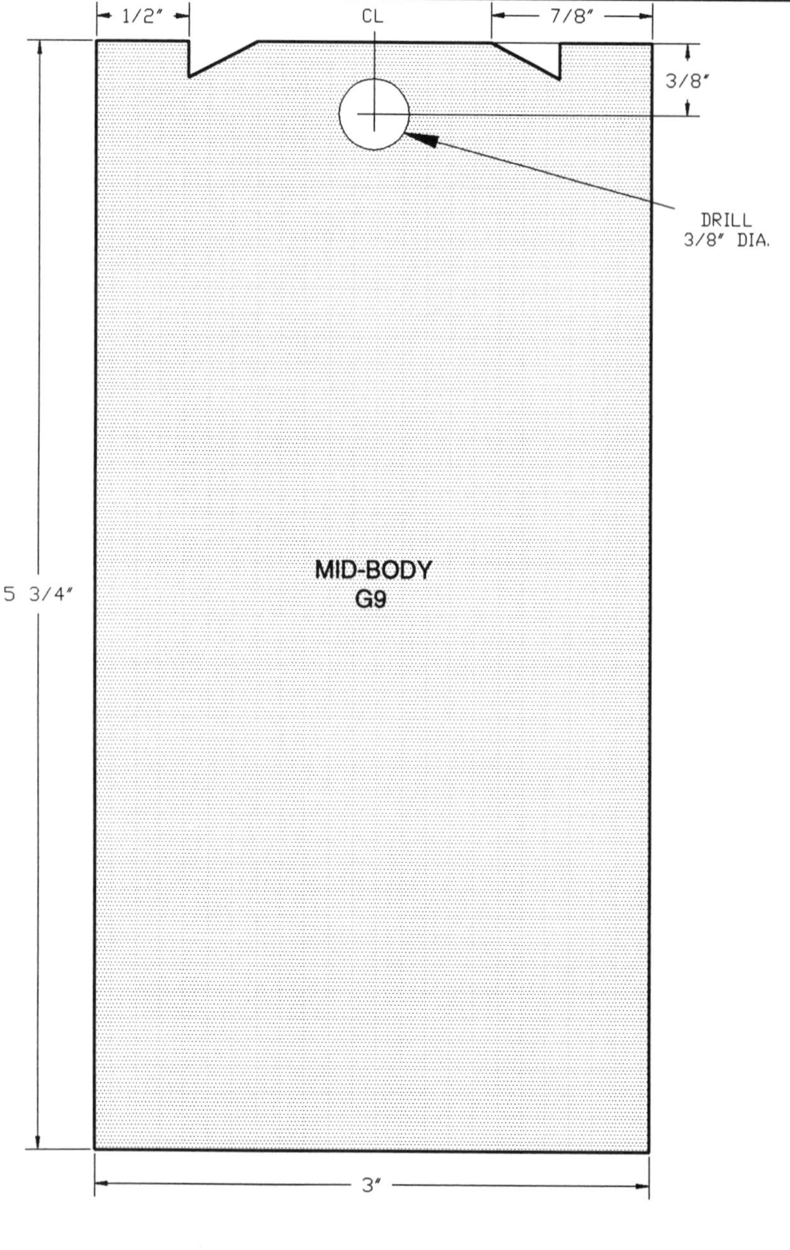
1/2″
CL
7/8″
3/8″
DRILL
3/8″ DIA.
5 3/4″
MID-BODY
G9
3″

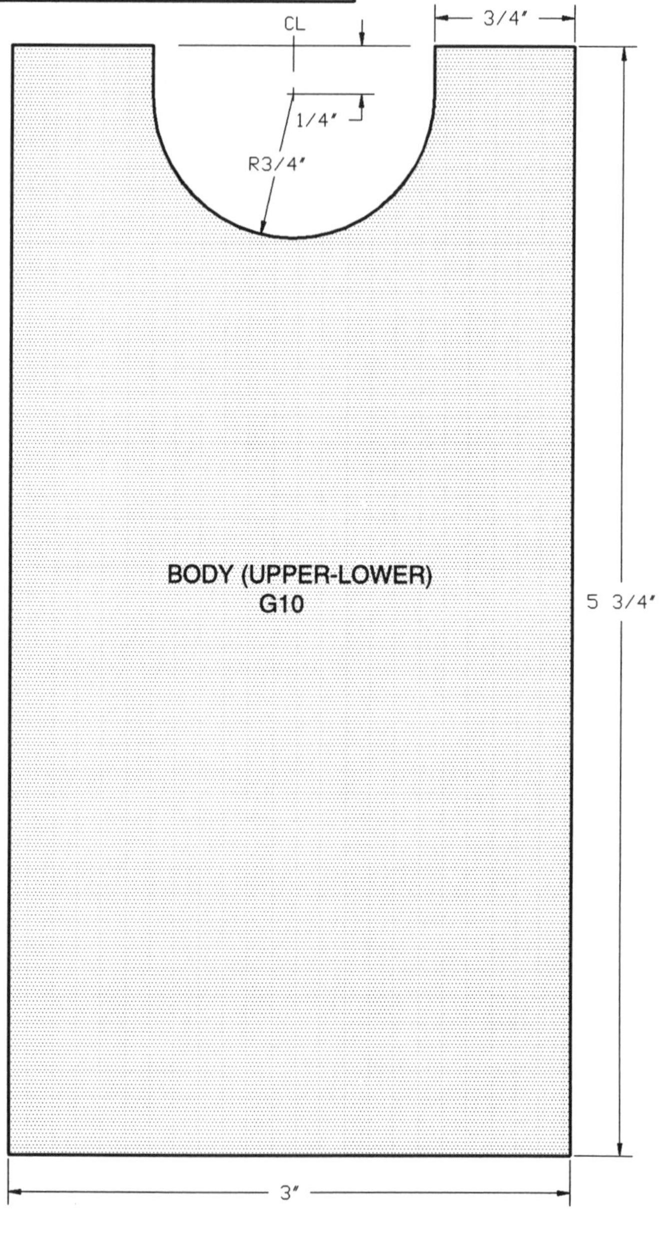
CL
3/4″
1/4″
R3/4″
BODY (UPPER-LOWER)
G10
5 3/4″
3″

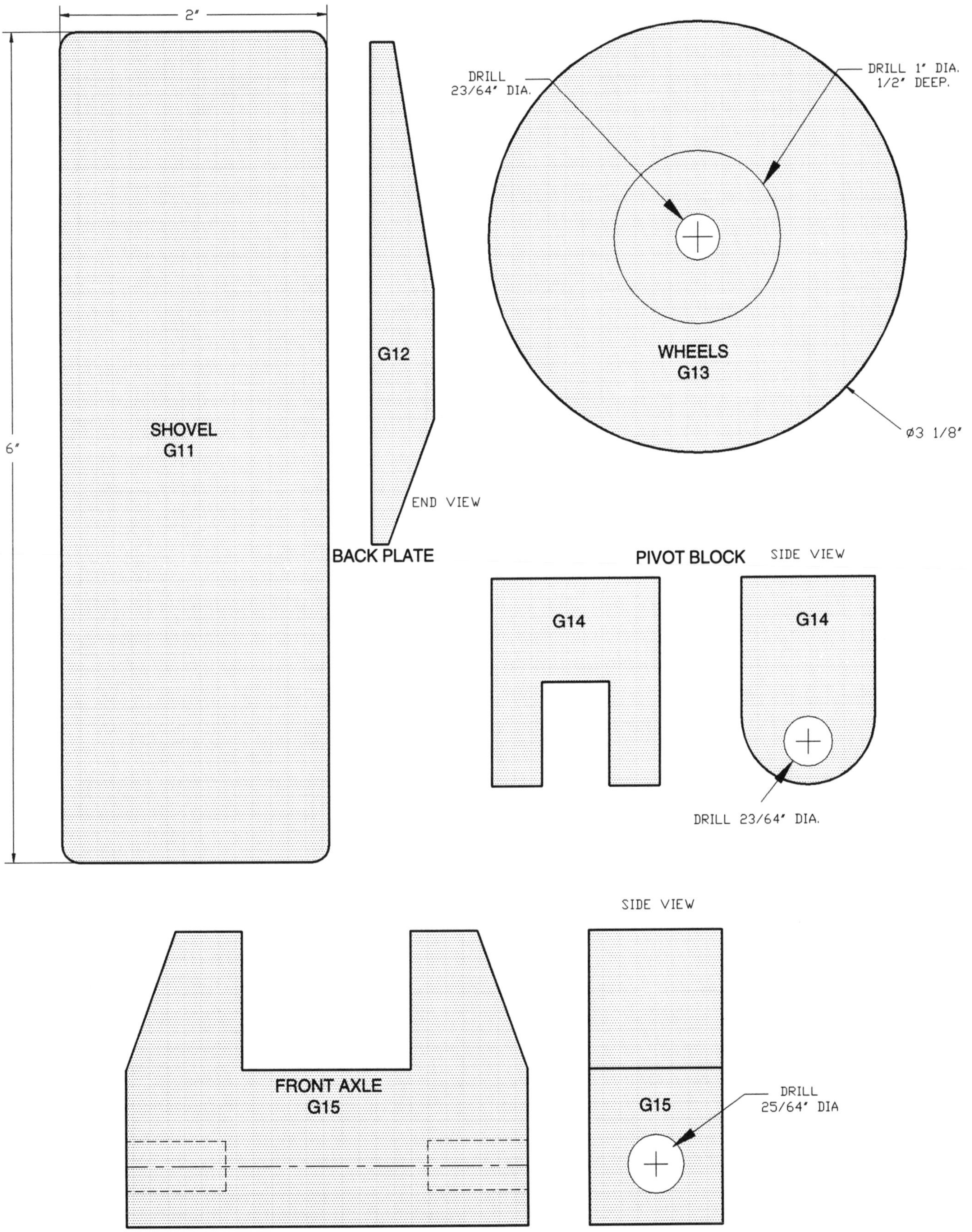
2"
6"
SHOVEL
G11
G12
END VIEW
BACK PLATE
DRILL
23/64" DIA.
DRILL 1" DIA.
1/2" DEEP.
WHEELS
G13
ø3 1/8"
PIVOT BLOCK
SIDE VIEW
G14
G14
DRILL 23/64" DIA.
SIDE VIEW
FRONT AXLE
G15
G15
DRILL
25/64" DIA

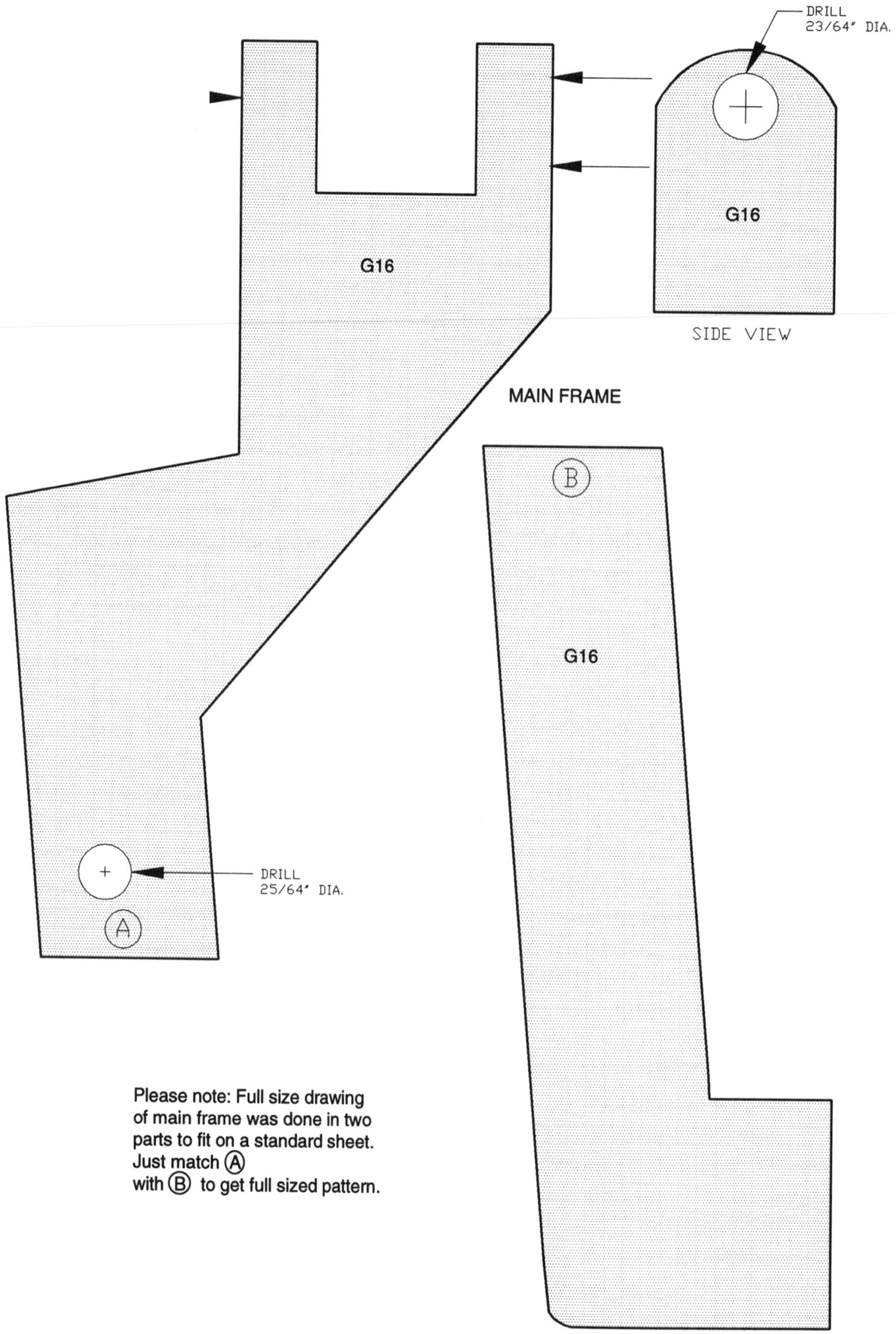

Please note: Full size drawing of main frame was done in two parts to fit on a standard sheet. Just match Ⓐ with Ⓑ to get full sized pattern.

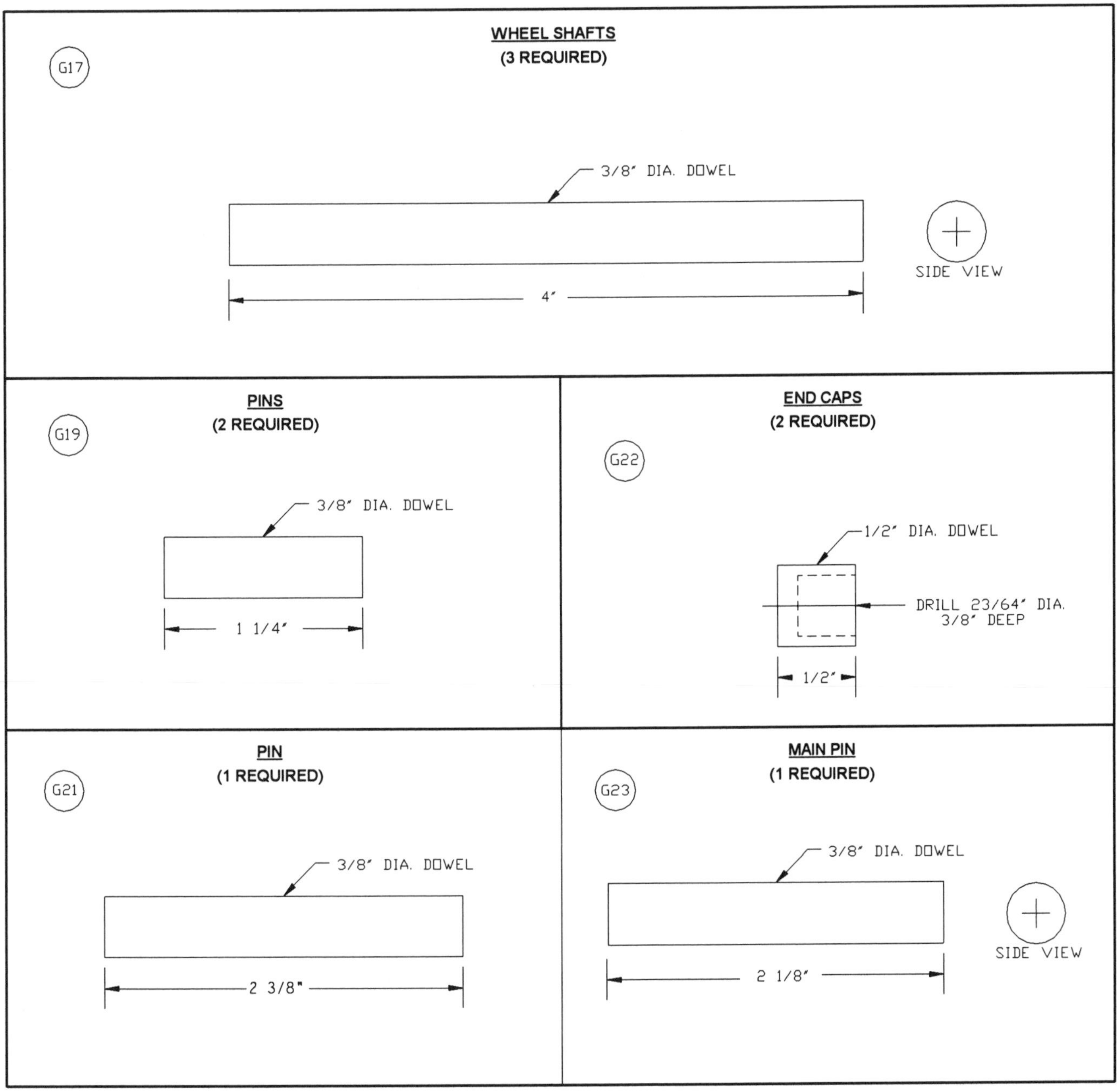
WHEEL SHAFTS
(3 REQUIRED)
G17
3/8" DIA. DOWEL
4"
SIDE VIEW
PINS
(2 REQUIRED)
G19
3/8" DIA. DOWEL
1 1/4"
END CAPS
(2 REQUIRED)
G22
1/2" DIA. DOWEL
DRILL 23/64" DIA.
3/8" DEEP
1/2"
PIN
(1 REQUIRED)
G21
3/8" DIA. DOWEL
2 3/8"
MAIN PIN
(1 REQUIRED)
G23
3/8" DIA. DOWEL
2 1/8"
SIDE VIEW

Additional Information - Grader

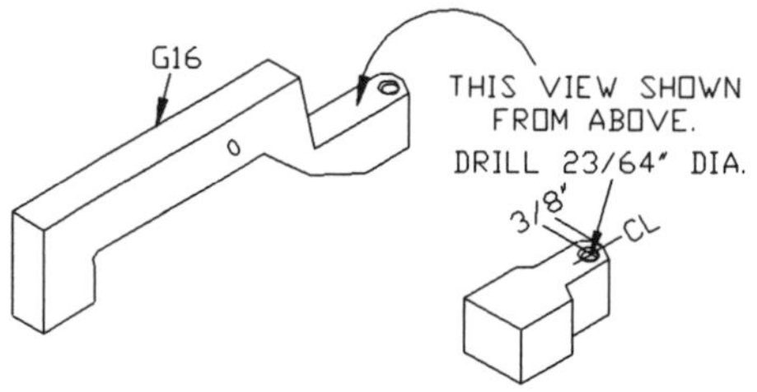

Drill 23/64" diameter hole in part G16.

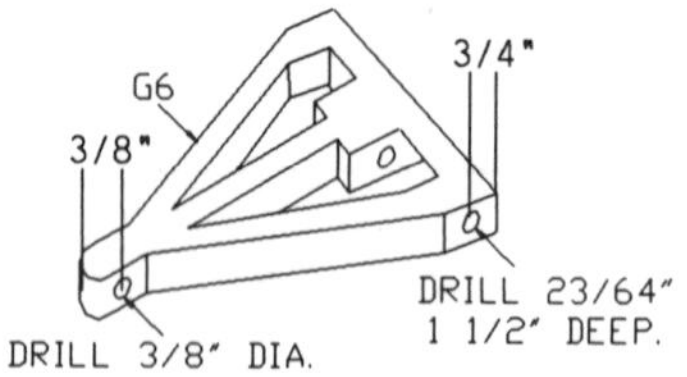

Drill holes in part G6, as shown.

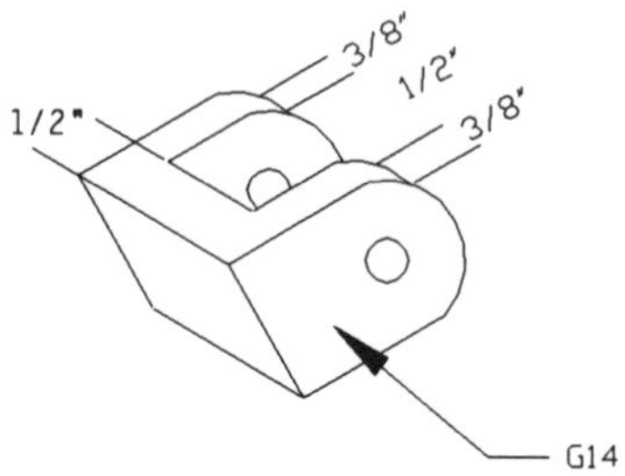

Cut groove in part G14, as shown.

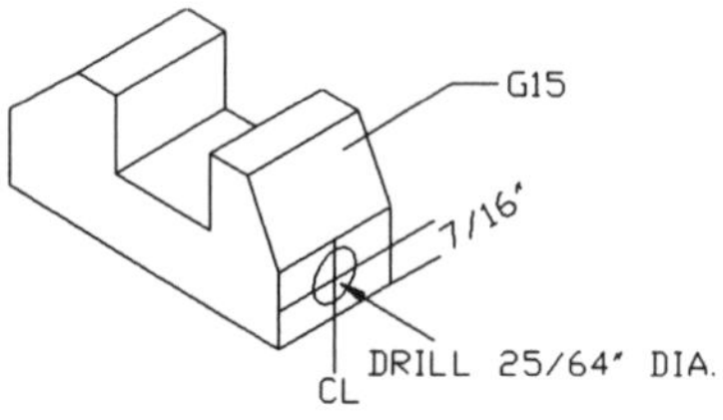

Drill 25/64" diameter hole in part G15.

Grader - Assembly Drawings

1

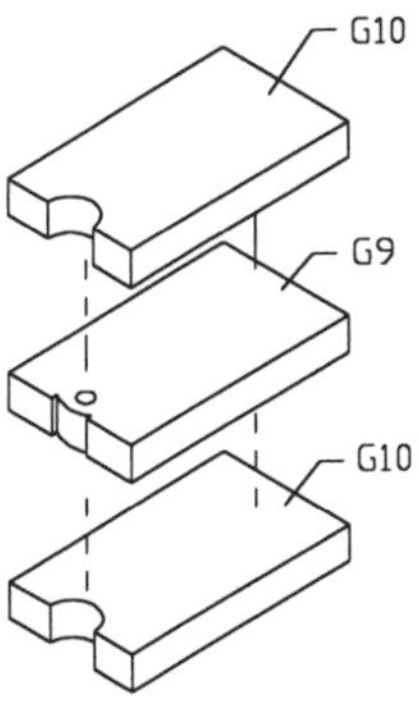

Glue mid-body G9 between upper and lower body G10.

2

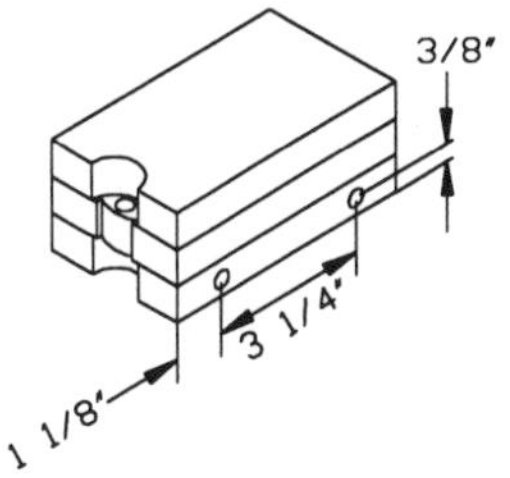

Trace hole location, as shown.

3

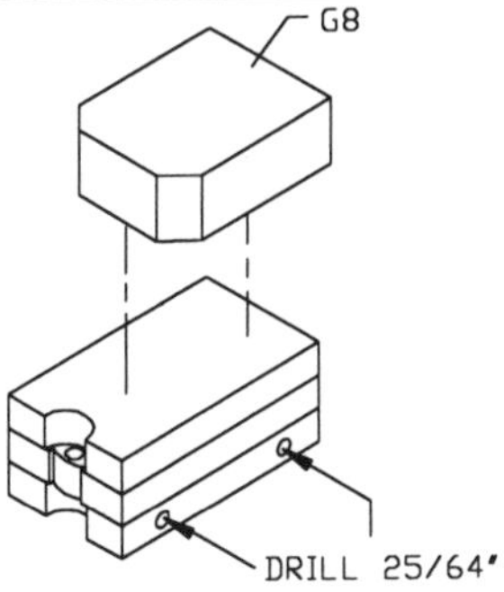

Drill 25/64" Dia. holes.
Glue rear hood G8.

4

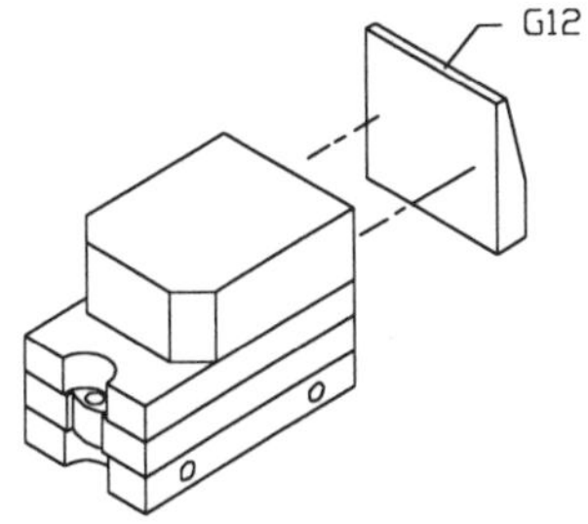

Glue rear plate G12.

5

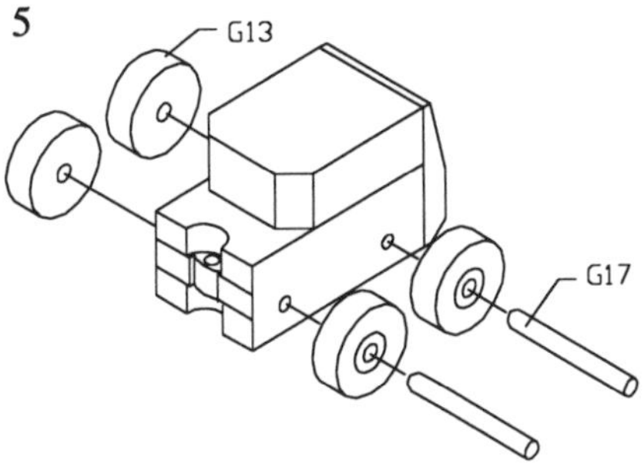

Insert wheel shafts G17 through wheels G13 and body.

6

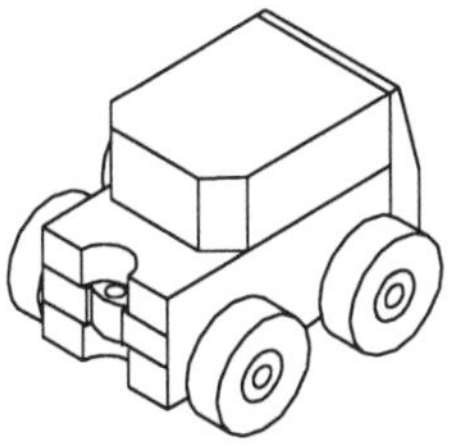

Start cabin assembly.

7

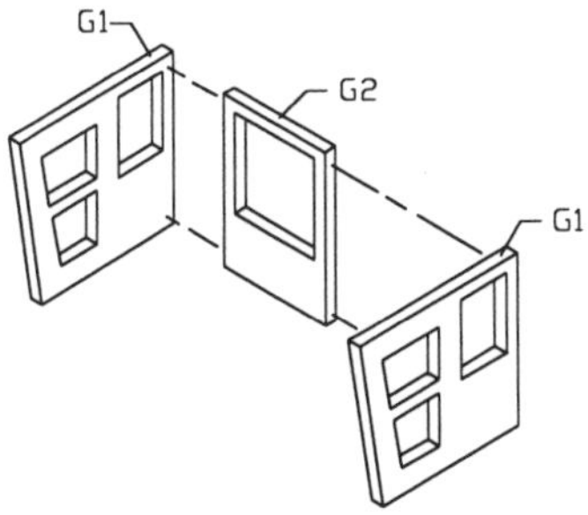

Glue cabin sides G1 to cabin back G2.

8

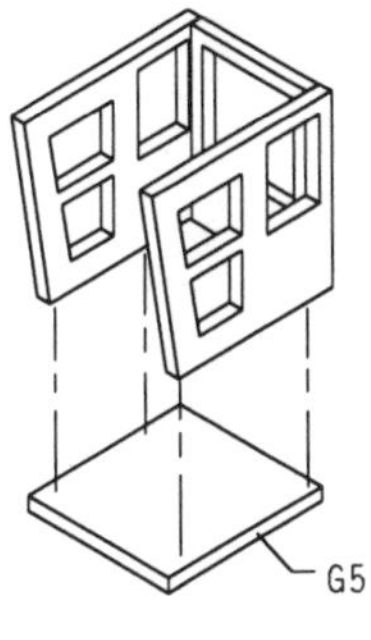

Glue cabin assembly onto floor G5.

9

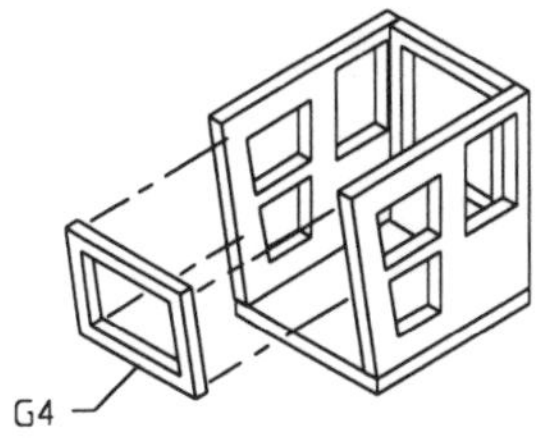

Glue cabin front G4.

10

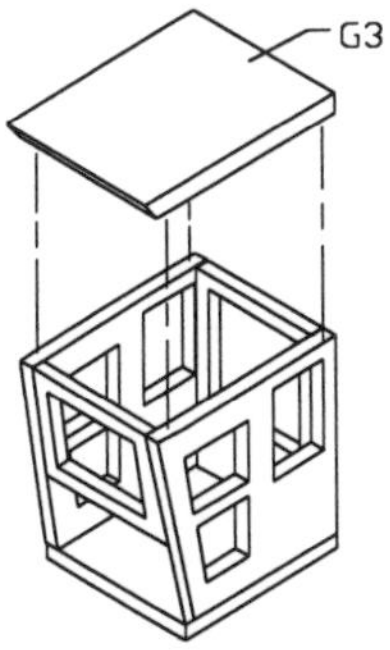

Glue roof G3 onto cabin.

11

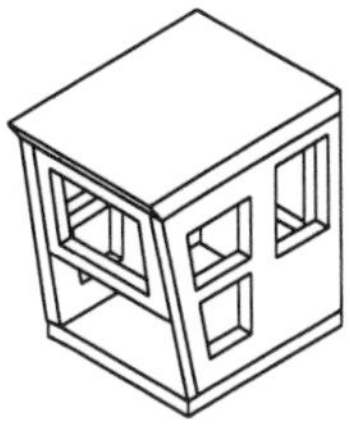

Your cabin should now be complete.

12

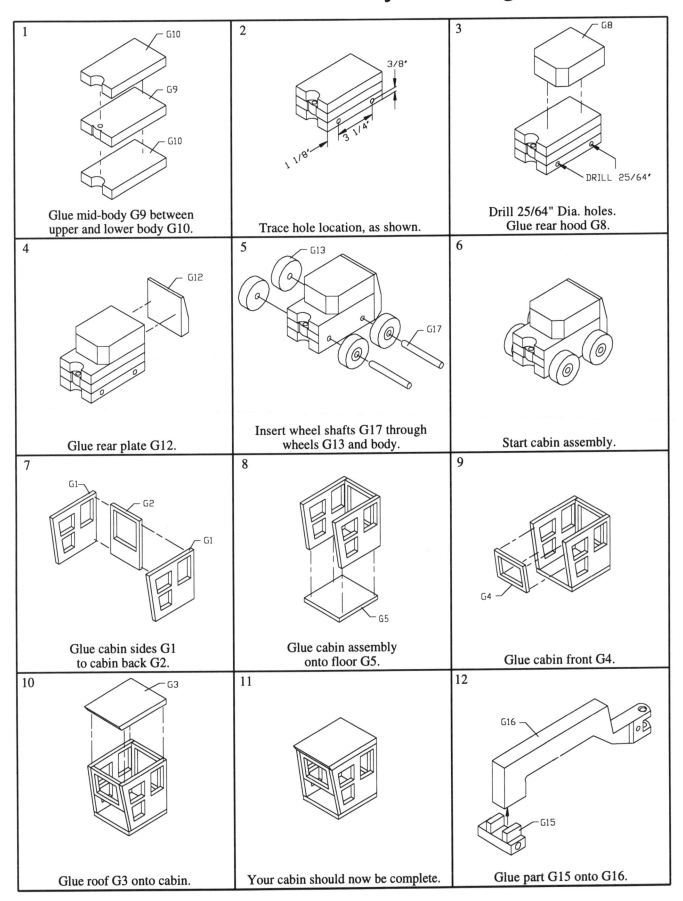

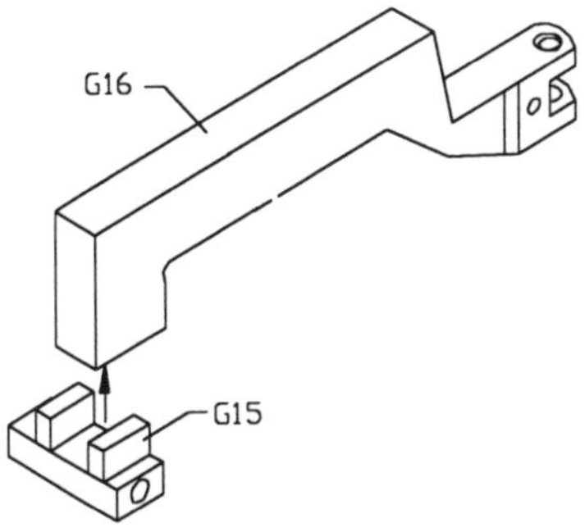

Glue part G15 onto G16.

13

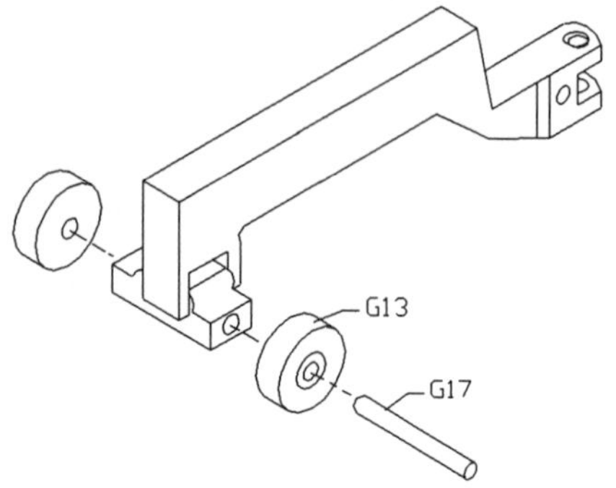

Insert wheel shaft G17 through Wheels G13 and into body.

14

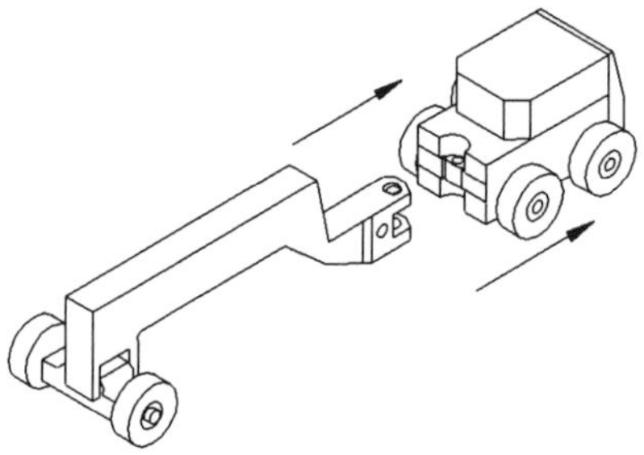

Slide front assembly into rear body.

15

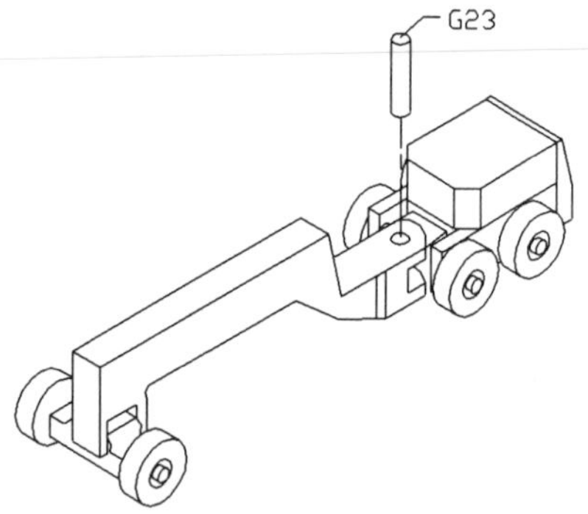

Insert main pin G23.

16

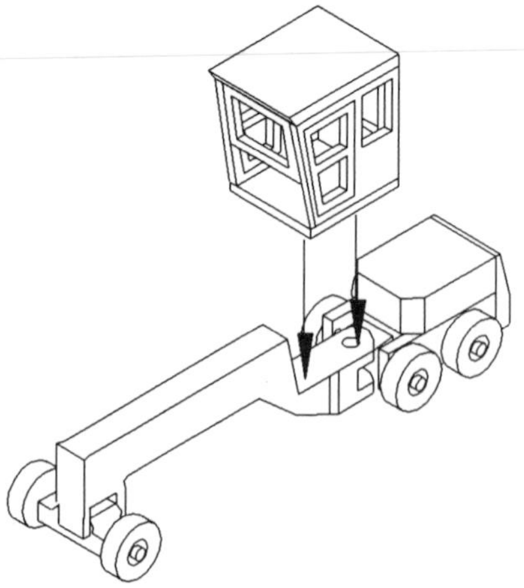

Glue cabin onto main frame.

17

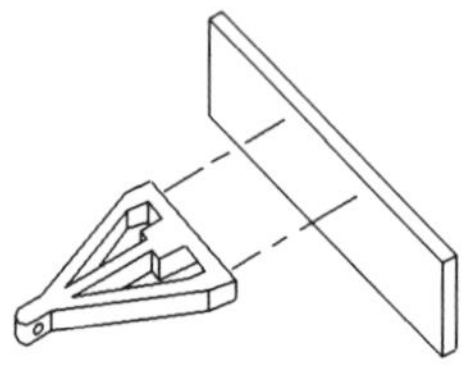

Glue shovel frame assembly G6 onto shovel, as shown.

18

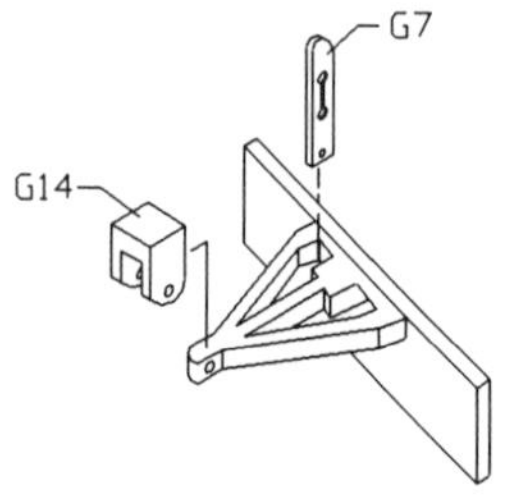

Insert control arm lever G7 and pivot block G14.

19

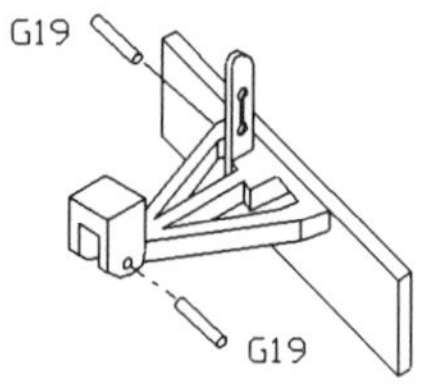

Insert pins G19.

20

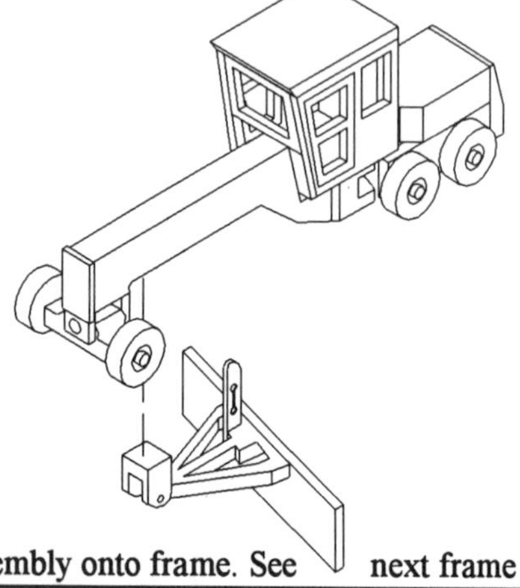

Glue assembly onto frame. See next frame for location.

21

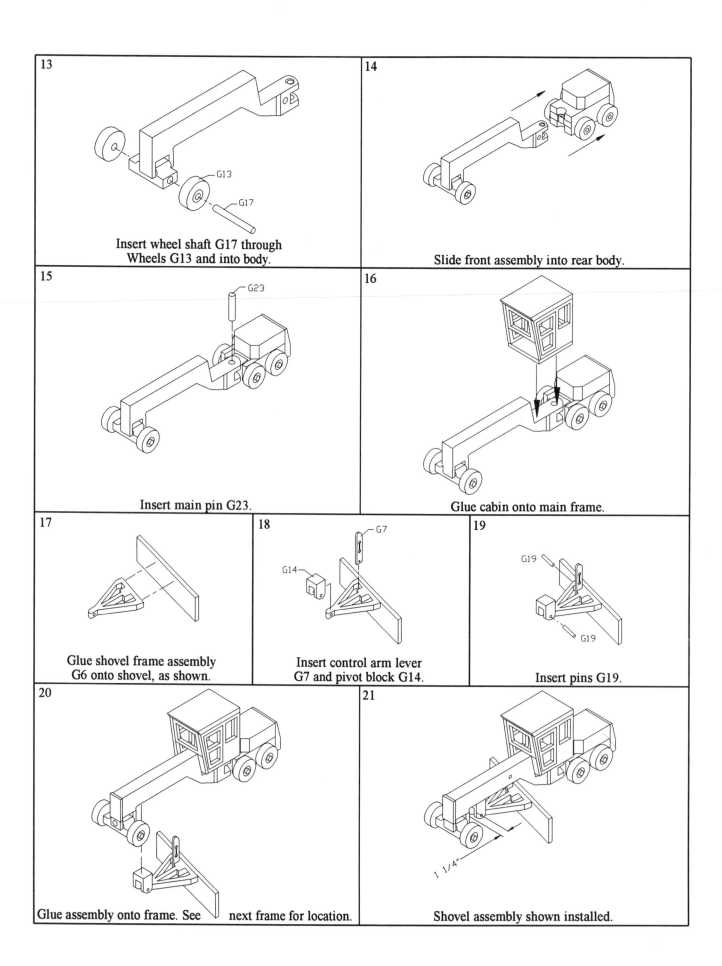

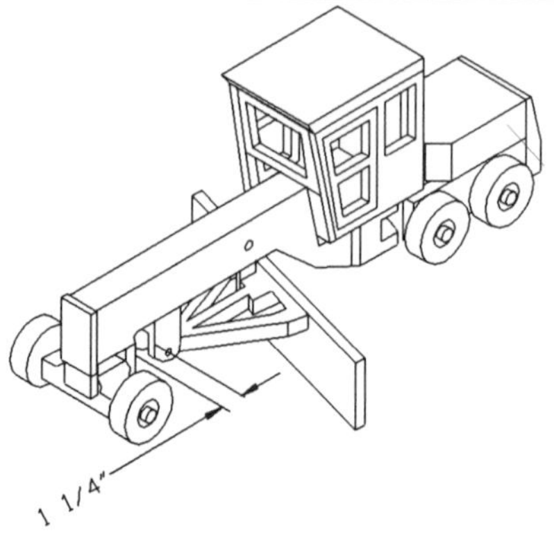

Shovel assembly shown installed.

22

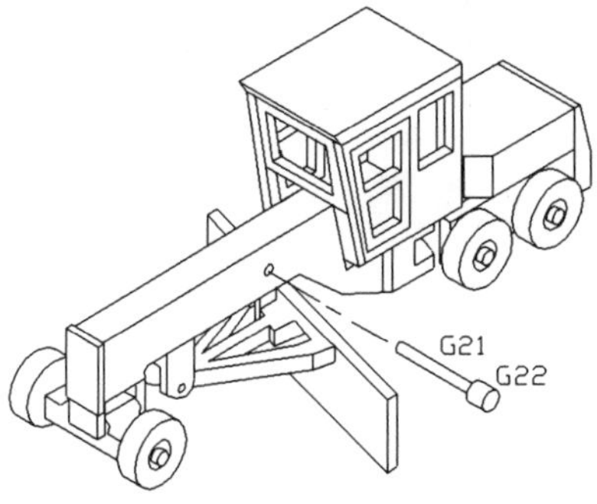

Glue together pin G21 and end cap G22.
Slide assembly into body. Do not glue.

23

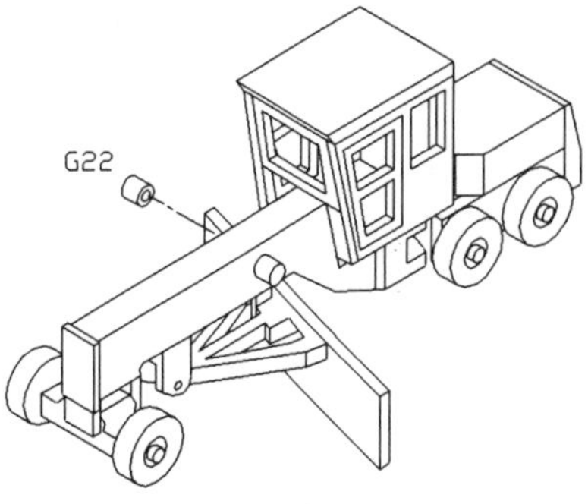

Glue other end cap G22.

24

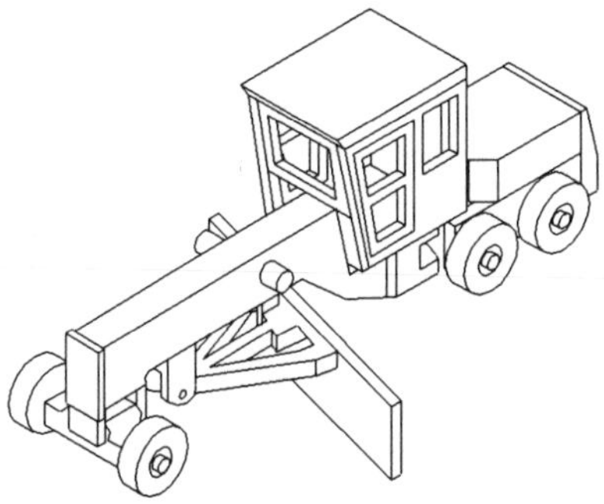

Slide the control pin from side to side to hold shovel in the up or down position.

25

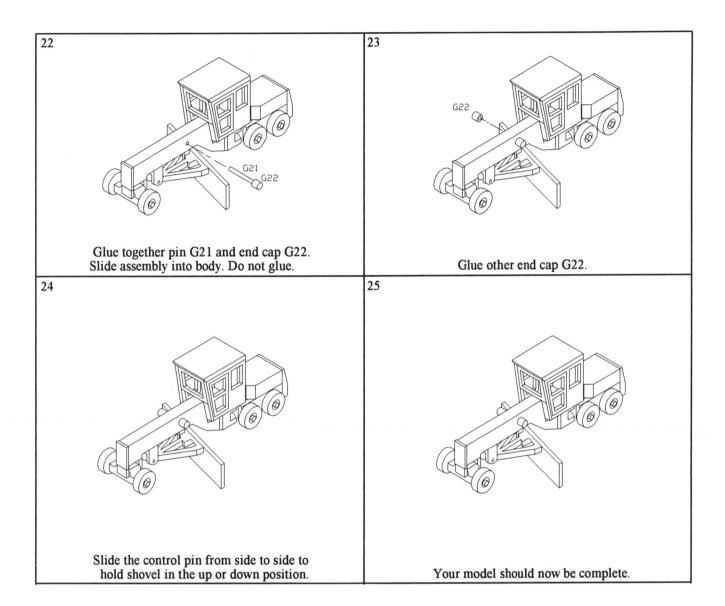

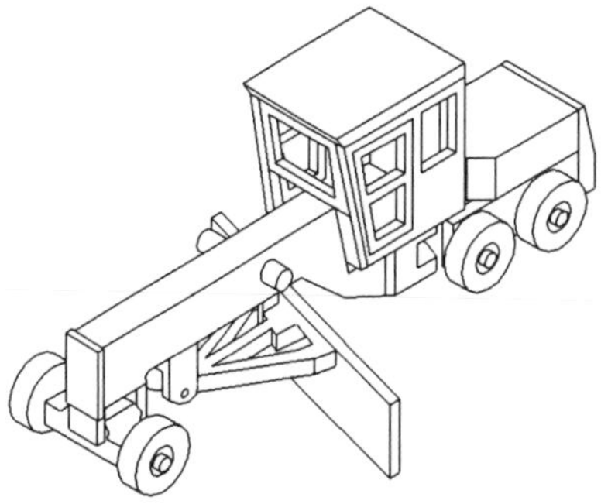

Your model should now be complete.

SKIDDER

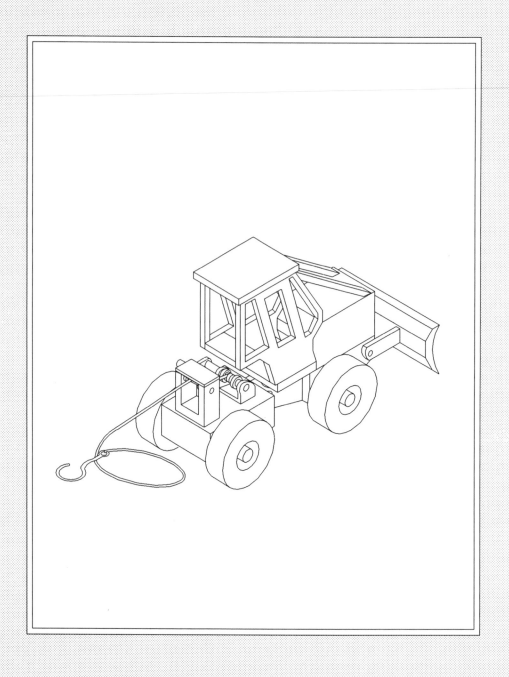

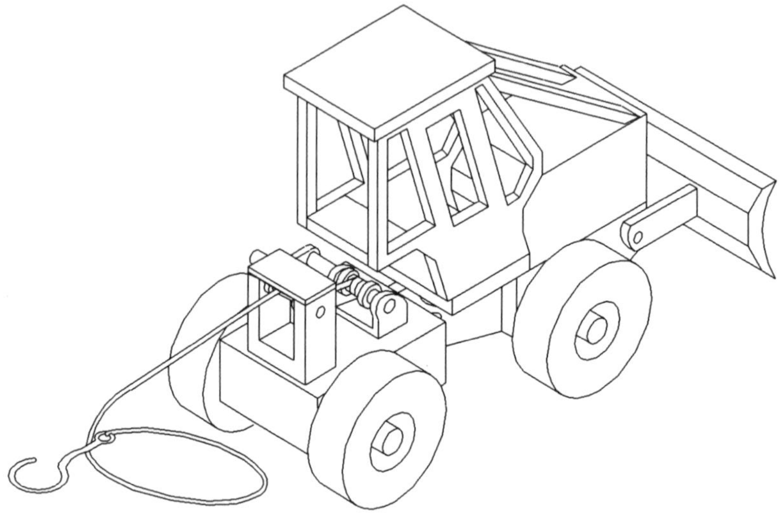

General Instructions - Skidder

1- Start by cutting materials needed by following the list of materials, paying attention to the rough and finished size. **Identify the parts as they are cut.**

Please note: Different types of wood can be used for the various parts. It is suggested, however, that hard wood be used, since many of the parts would be much too fragile if using soft wood. We have used a combination of pine, maple and oak to give the models a nice contrast!

2- Remove the full-size patterns found in the appendix. Cut them out, leaving approximately 1/16" all around, and place on the proper piece of wood. Patterns can be secured to wood using either spray adhesive or rubber ciment. If using the latter, cut and sand the part first to finished size. If drilling is required, mark the hole by inserting a scriber or nail through the pattern into the wood. Remove the pattern before drilling.

You should have no trouble determining which surface to attach most of the patterns. Some parts, however, can be confusing since the pattern could fit on more than one surface. The drawings below indicate exactly which surface to attach the patterns for these parts.

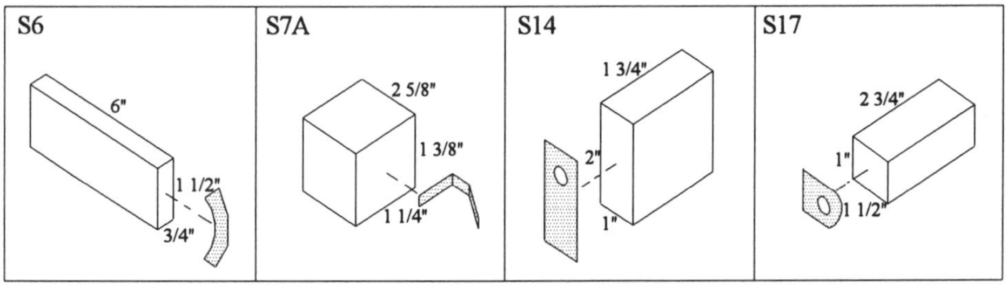

3- Look at the full-size drawing sheets to finish parts S2 and S16.

4- Parts S14 and S17 will need additional cuts and details, please refer to the additional information pages, to complete these parts.

5- Using maple dowels, make all pins, shafts, etc.

6- Follow the assembly drawings to complete your model.

List of Materials - Skidder

Part	T	W	L	Material	Qty.	*
S1	3/8"	3"	6 7/8"	pine	2	R
S2	3/4"	2 5/8"	5 1/2"	pine	1	F
S3	3/8"	1"	3"	maple	2	R
S4	1/2"	5/8"	3 3/8"	maple	2	F
S5	3/8"	1 1/2"	5 1/4"	pine	2	F
S6	3/4"	1 1/2"	6"	pine	1	F
S7	1 1/2"	3 3/8"	3 3/4"	pine	1	F
S7A	1 1/4"	1 3/8"	2 5/8"	pine	1	F
S8	3/4"	1 5/8"	2 3/4"	maple	1	R
S9	1/4"	3 3/8"	2 5/8"	pine	1	F

Part	T	W	L	Material	Qty.	*
S10	3/8"	1/2"	4 5/8"	oak	2	R
S11	1 3/4"	3 3/8"	3 15/16"	pine	1	F
S12	3/8"	3 3/4"	3 1/8"	pine	1	F
S13	3/8"	4"	4"	oak	2	R
S14	1"	2"	1 3/4"	maple	1	F
S15	1/4"	1"	1 3/4"	maple	1	F
S16	1 1/4"	3 3/4" DIA.		oak	4	F
S17	1"	1 1/2"	2 3/4"	maple	1	F
S18	1/4"	1 1/4"	2"	maple	1	R
S19	3/8"	1 1/8"	2"	maple	1	R

R = Rough size
F = Finished size

T = Thickness
W = Width
L = Length

Instructions:

R= Rough sizes, the material is cut oversized so you have ample room to apply the pattern on the surface. Sanding is not required at this point.

F = Finished Size: Cut and sand parts to finished size.

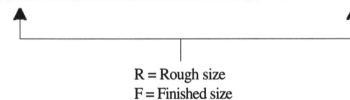

Full-Sized Patterns: Set One

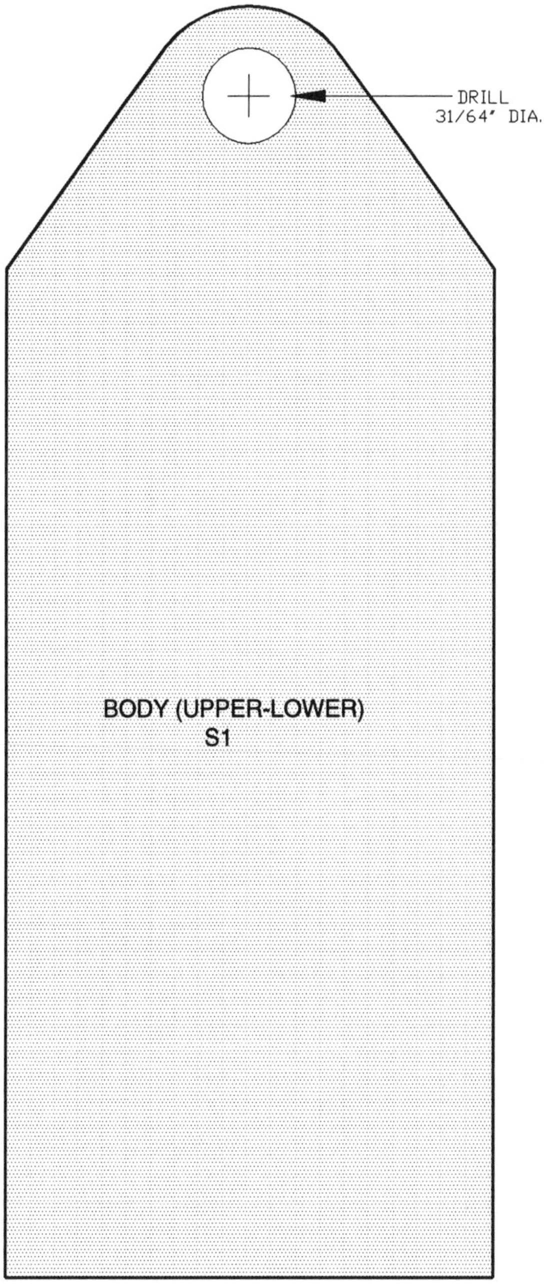

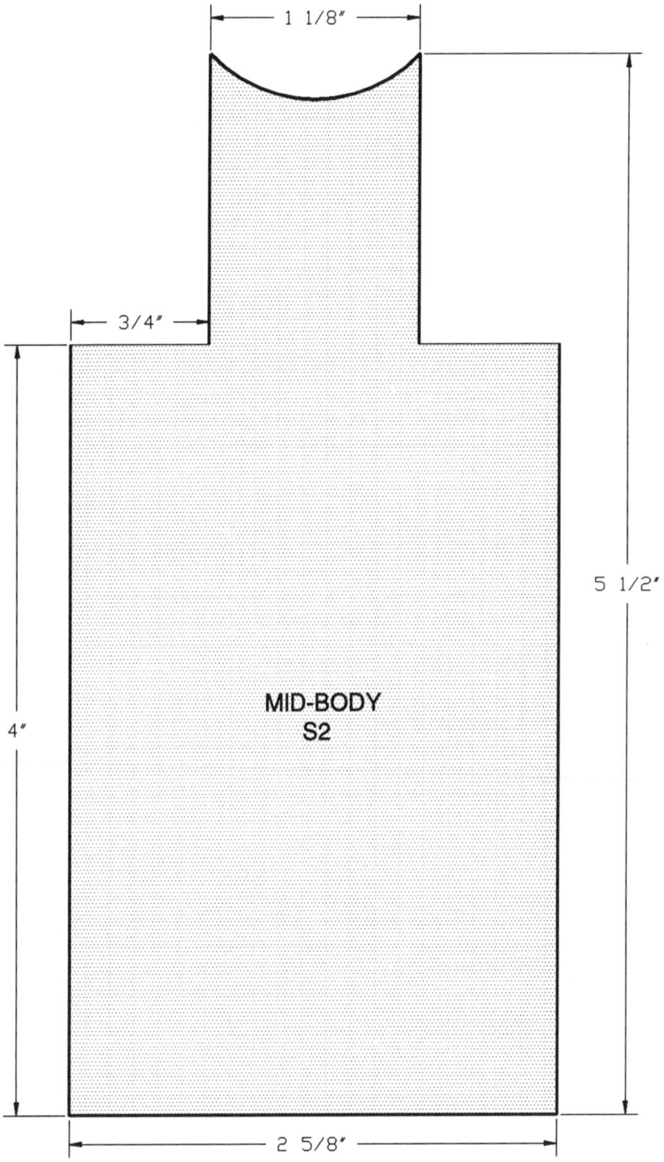

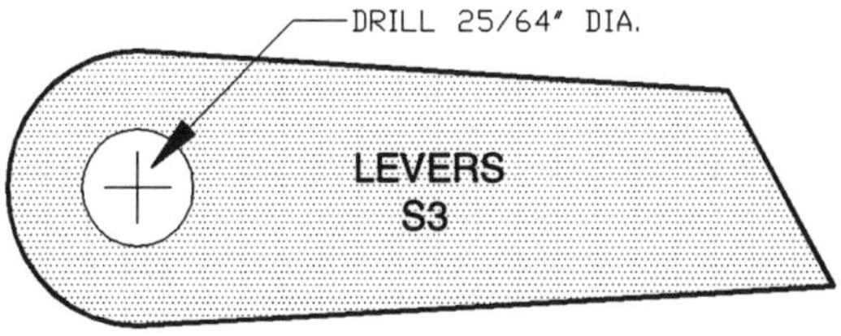

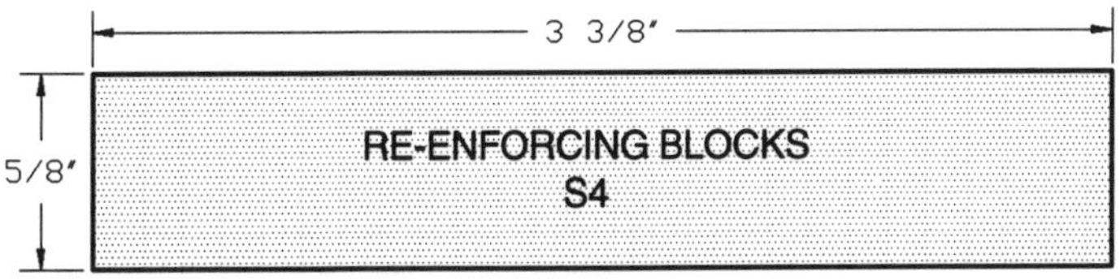

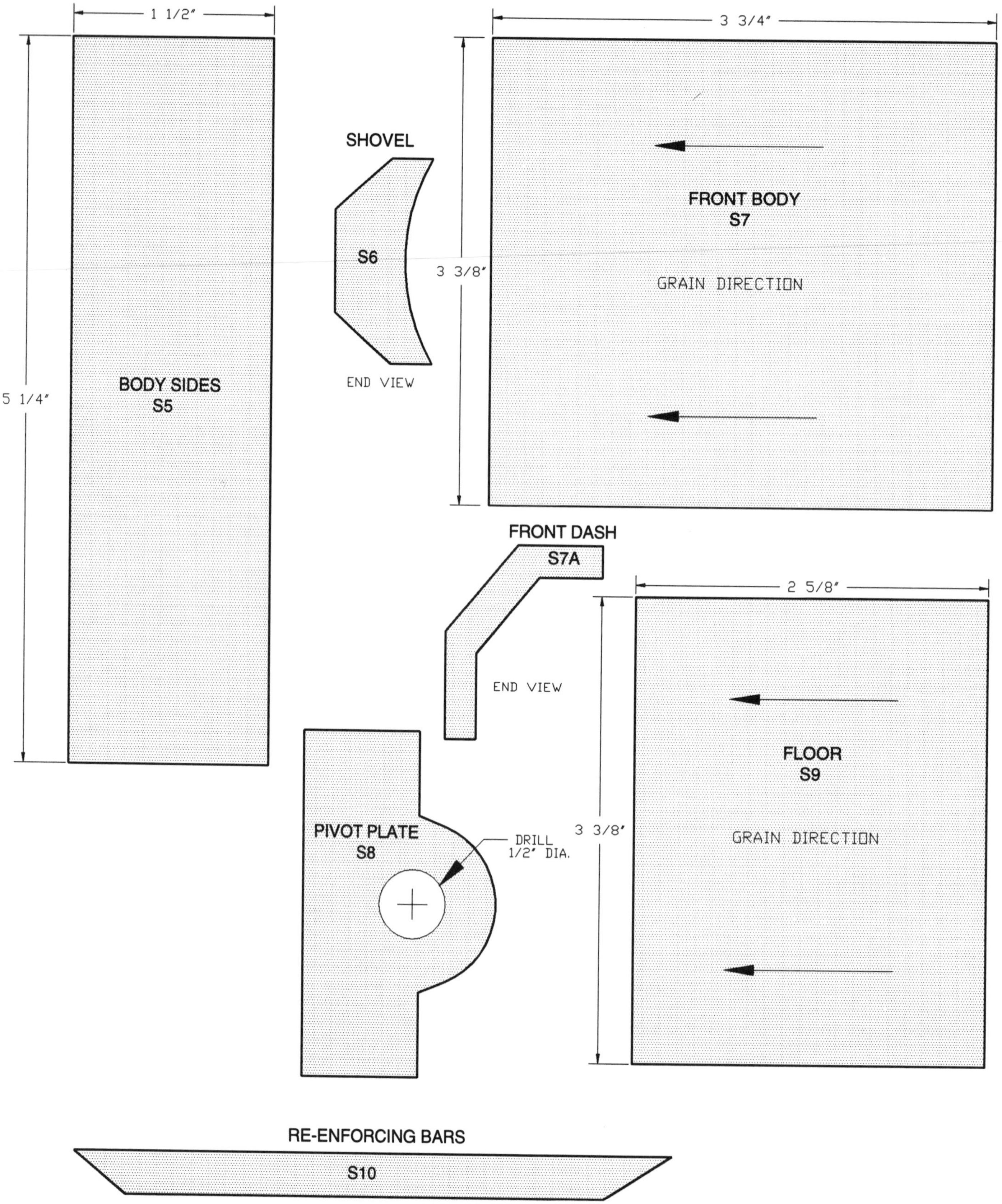
1 1/2"
5 1/4"
BODY SIDES
S5
SHOVEL
S6
END VIEW
3 3/4"
3 3/8"
FRONT BODY
S7
GRAIN DIRECTION
FRONT DASH
S7A
END VIEW
2 5/8"
PIVOT PLATE
S8
DRILL
1/2" DIA.
3 3/8"
FLOOR
S9
GRAIN DIRECTION
RE-ENFORCING BARS
S10

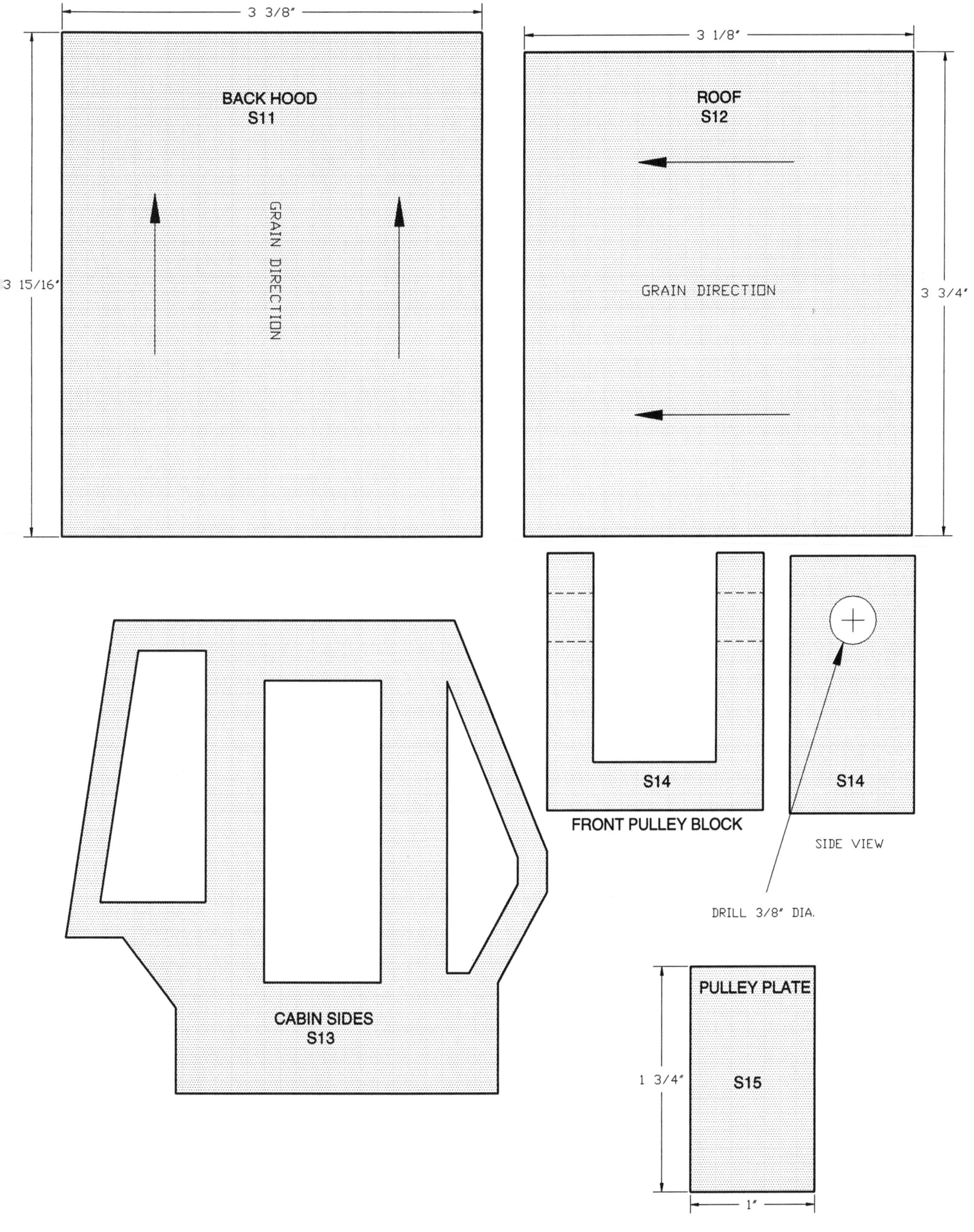
3 3/8"
3 15/16"
BACK HOOD
S11
GRAIN DIRECTION
3 1/8"
ROOF
S12
GRAIN DIRECTION
3 3/4"
CABIN SIDES
S13
S14
FRONT PULLEY BLOCK
S14
SIDE VIEW
DRILL 3/8" DIA.
PULLEY PLATE
S15
1 3/4"
1"

WHEELS

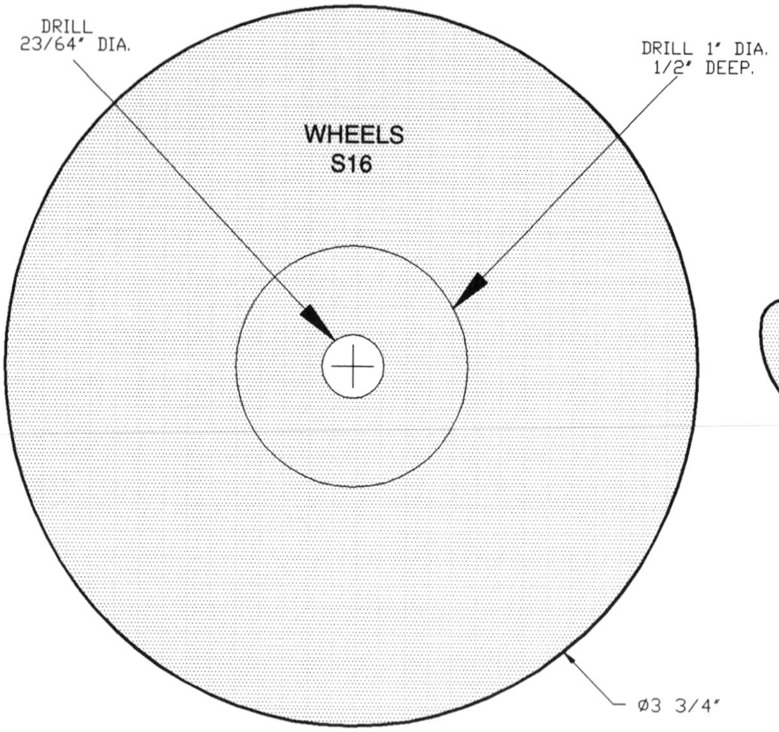

HOOK

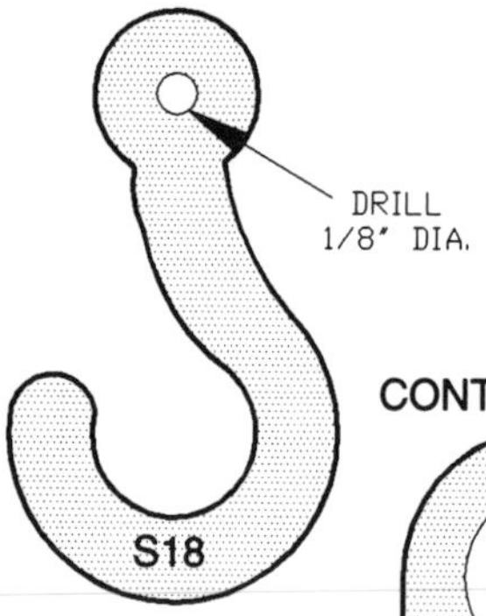

CONTROL LEVER

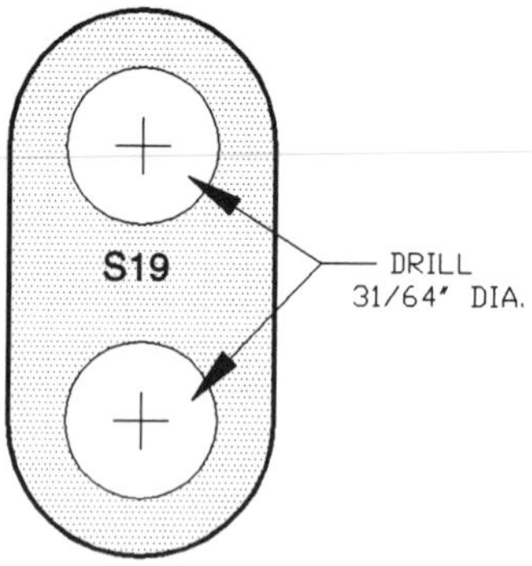

PULLEY FRAME

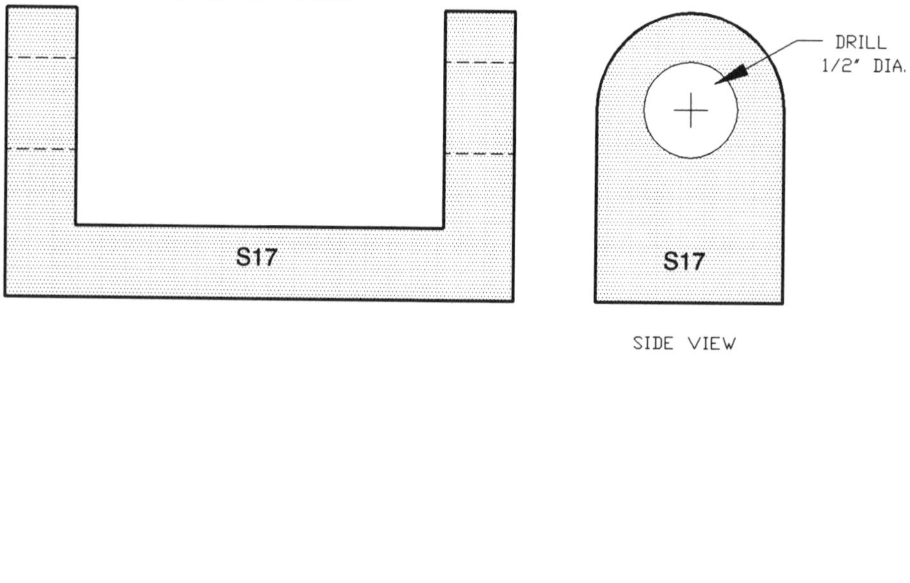

SIDE VIEW

WHEEL SHAFTS
(2 REQUIRED)

S20

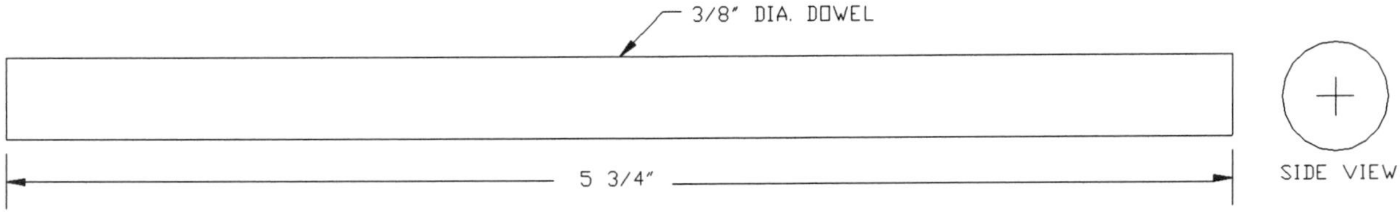

PULLEY SHAFT
(1 REQUIRED)

S21

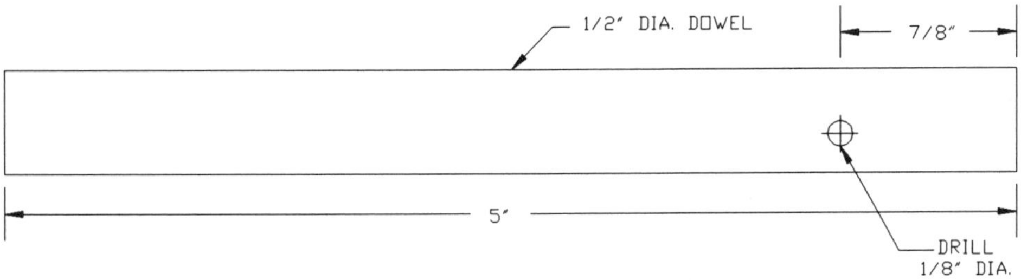

PULLEY SIDES
(2 REQUIRED)

S22

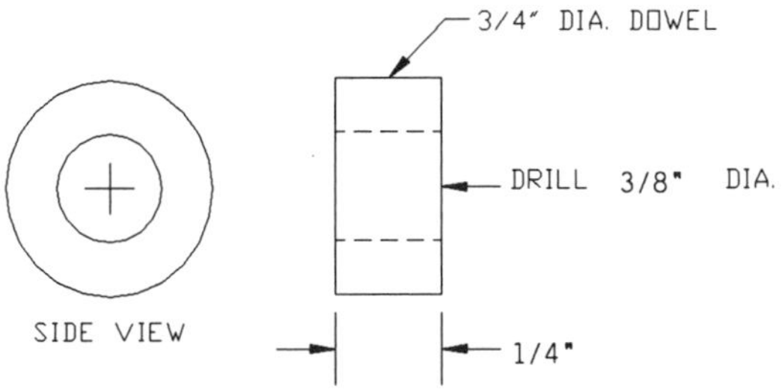

MAIN PIVOT PIN
(1 REQUIRED)

S23

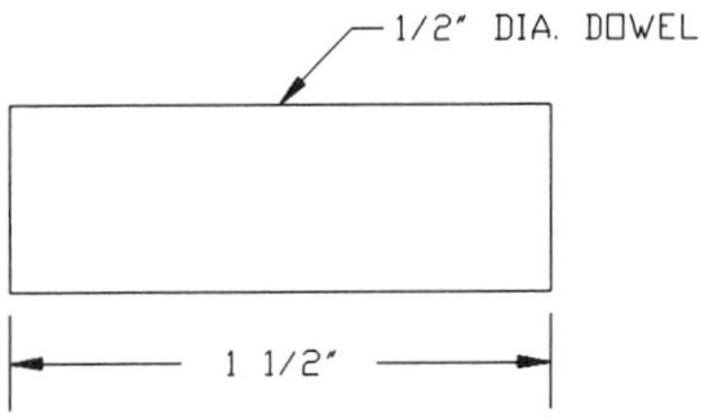

PULLEY SIDES
(2 REQUIRED)

S24

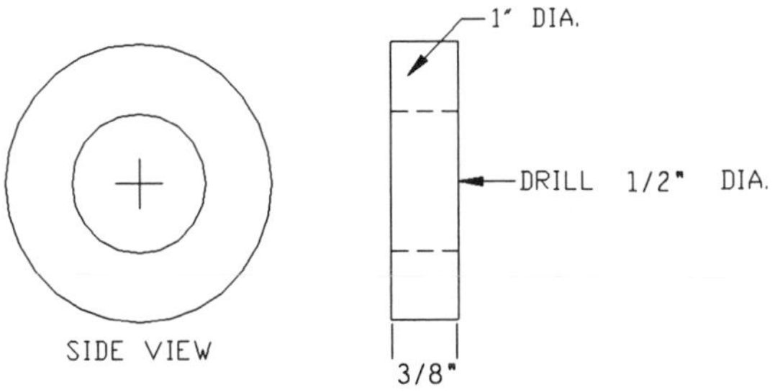

FRONT PULLEY SHAFT
(1 REQUIRED)

S25

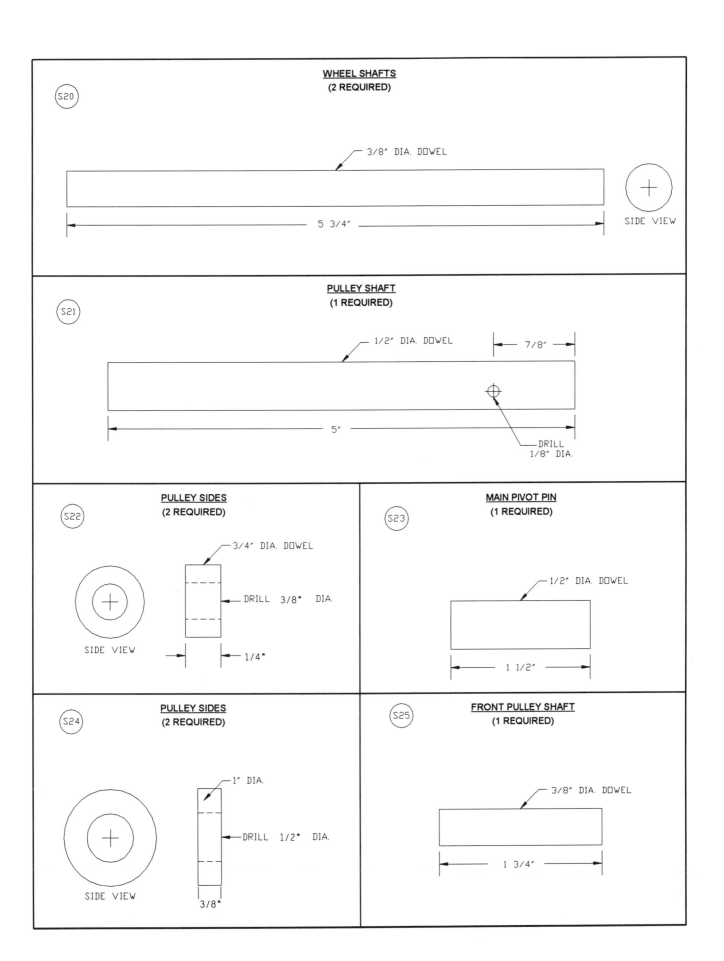

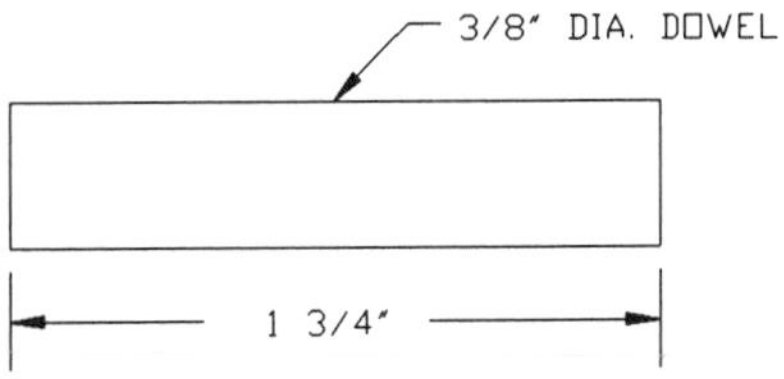

PINS
(2 REQUIRED)

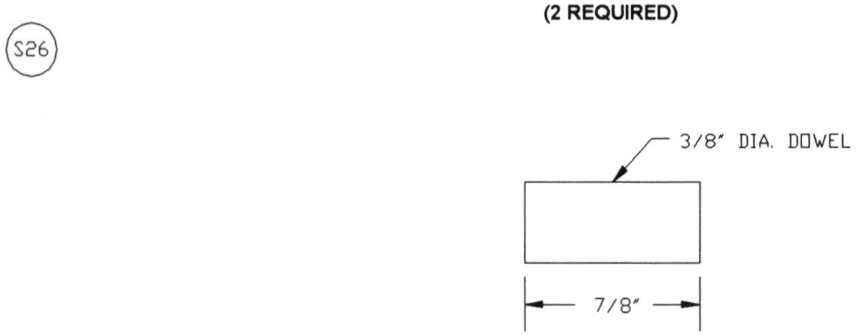

HANDLE
(1 REQUIRED)

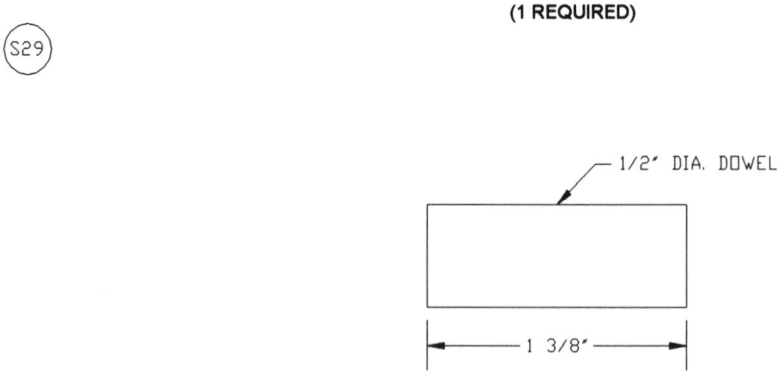

Additional Information - Skidder

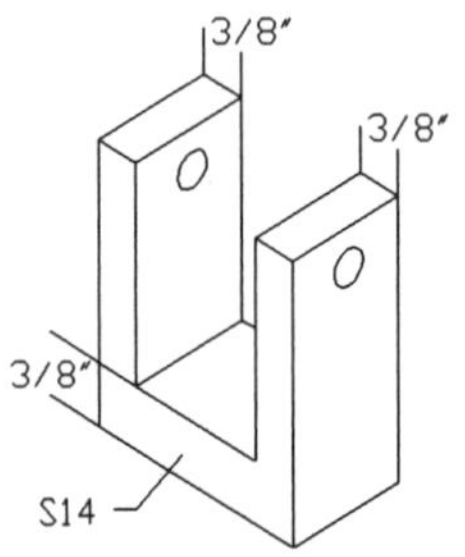

Using your Scroll Saw, cut out part S14, as shown.

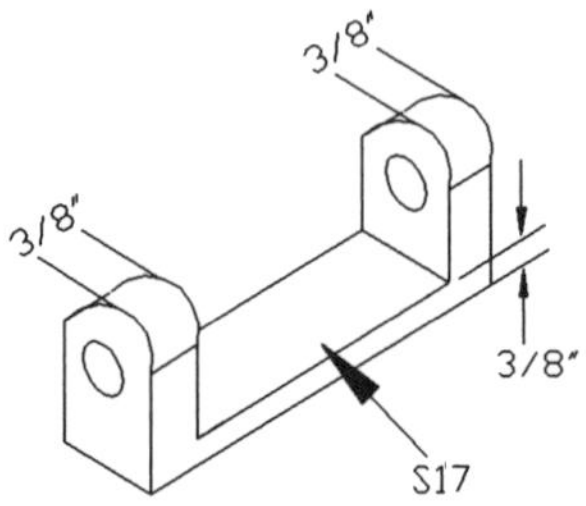

Using your Scroll Saw, cut out part S17, as shown.

Skidder - Assembly Drawings

1

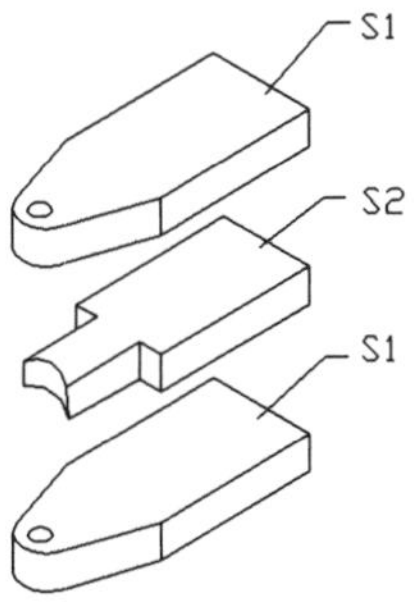

Glue mid-body S2 between upper and lower body S1.

2

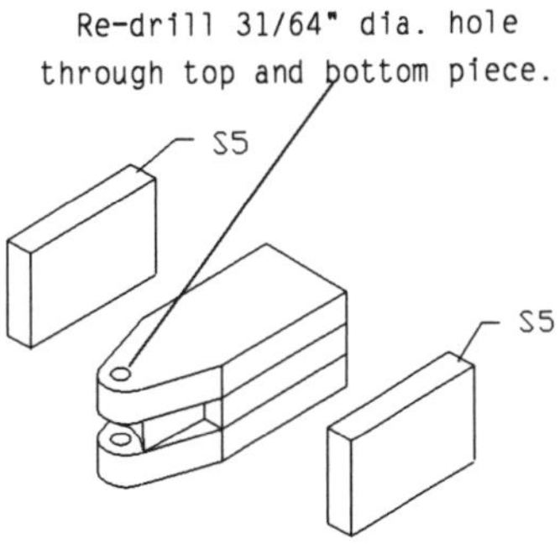

Re-drill hole, as shown. Glue sides S5.

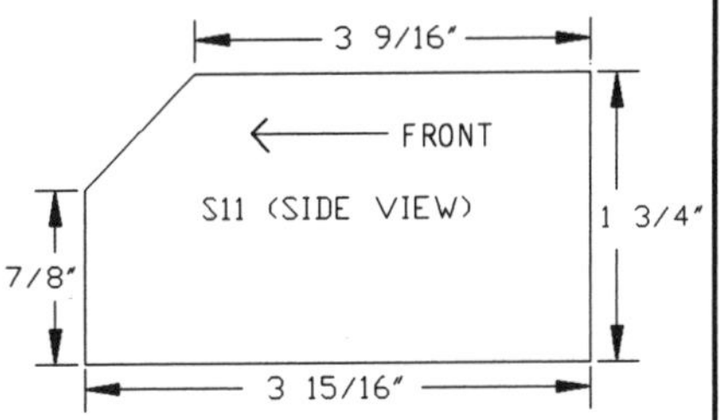

Sand angle on rear hood S11, as shown.

3

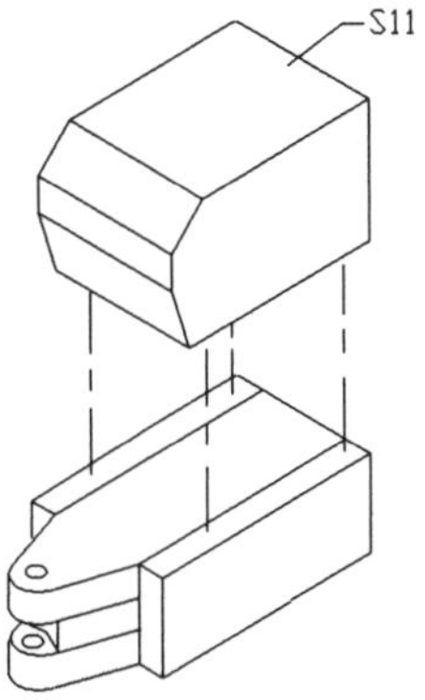

Glue rear hood S11 onto body.

4

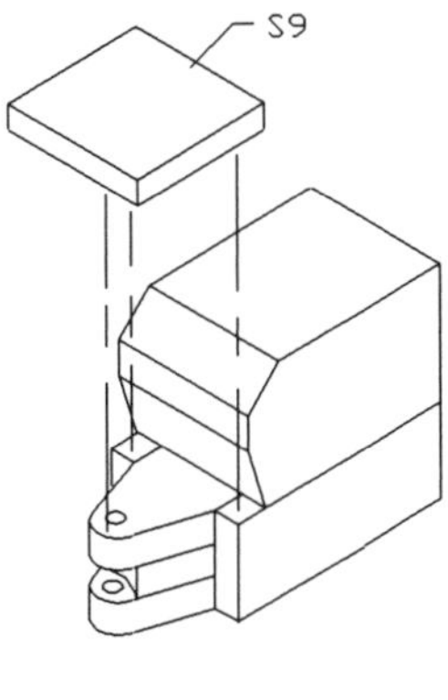

Glue cabin floor S9.

5

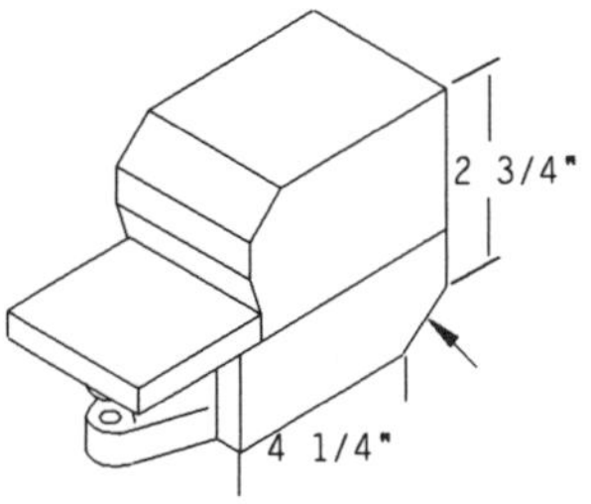

Sand angle (from side to side).

6

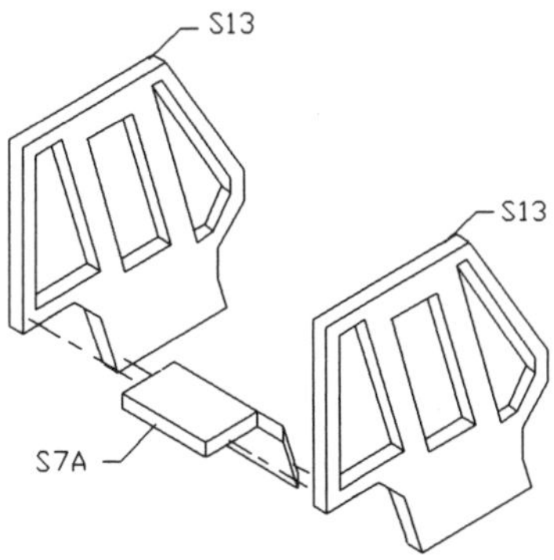

Glue front dash S7A, as shown.

7

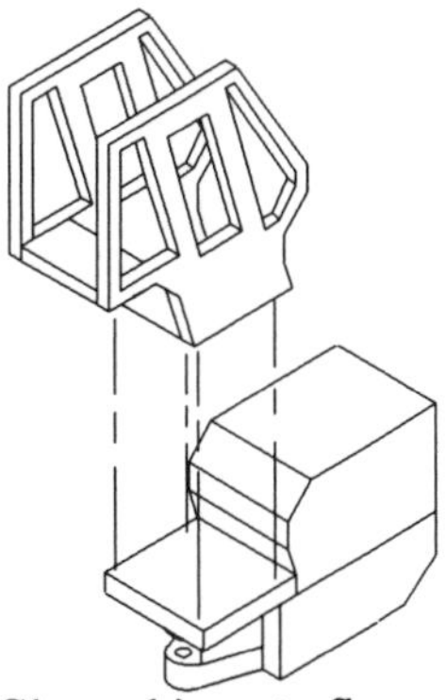

Glue cabin onto floor.

8

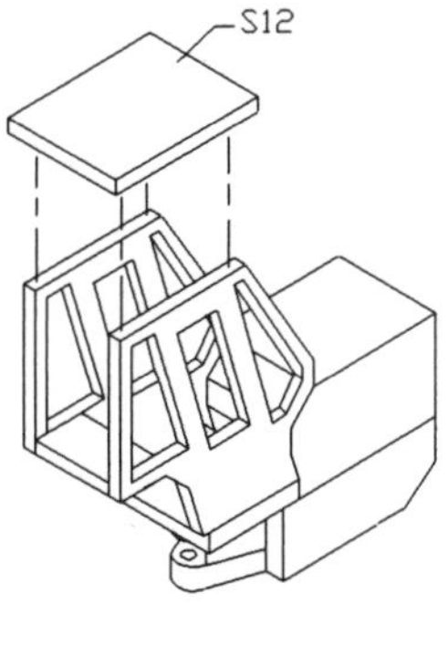

Glue roof S12 onto cabin.

9

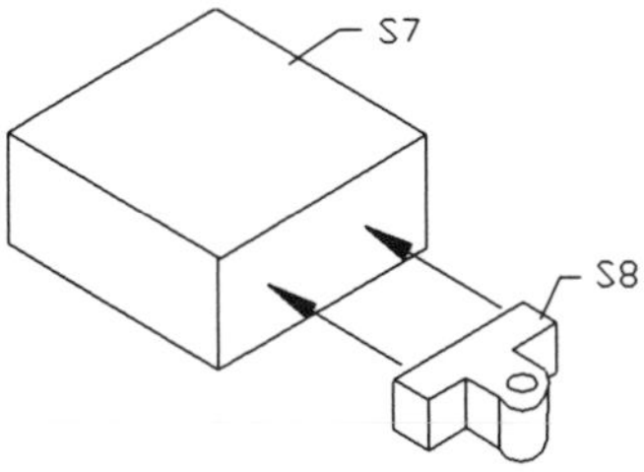

Glue pivot plate S8 onto body S7. See next frame for exact location.

10

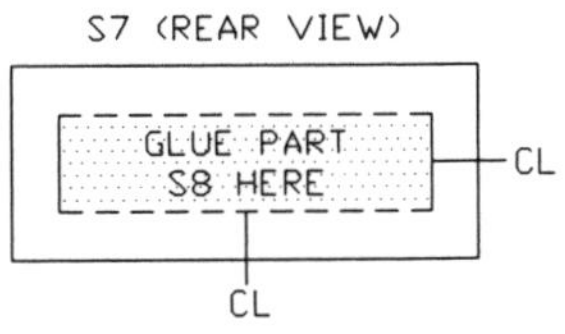

Where to glue pivot plate S8 onto body S7.

11

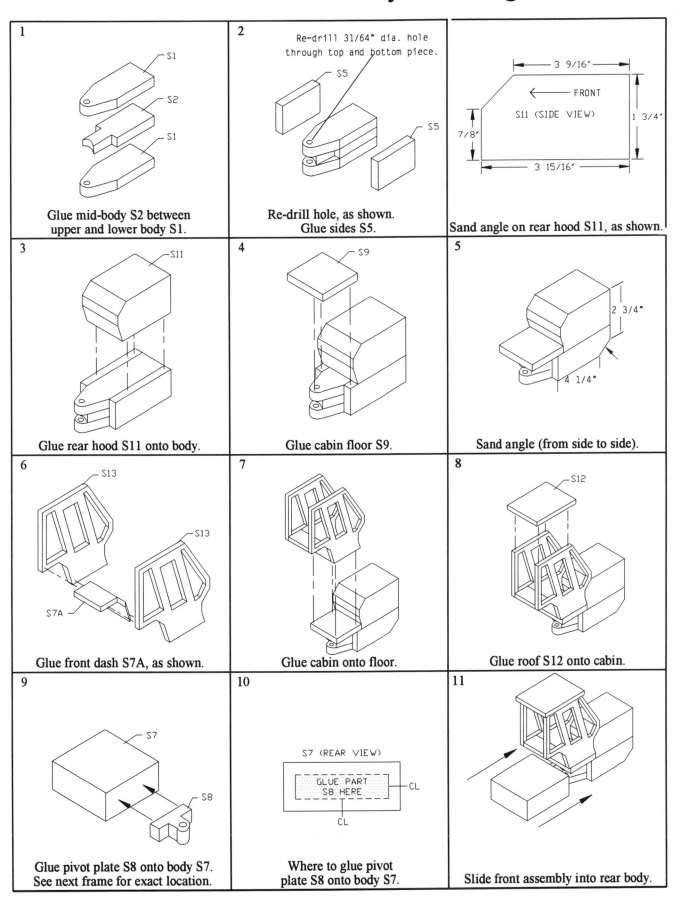

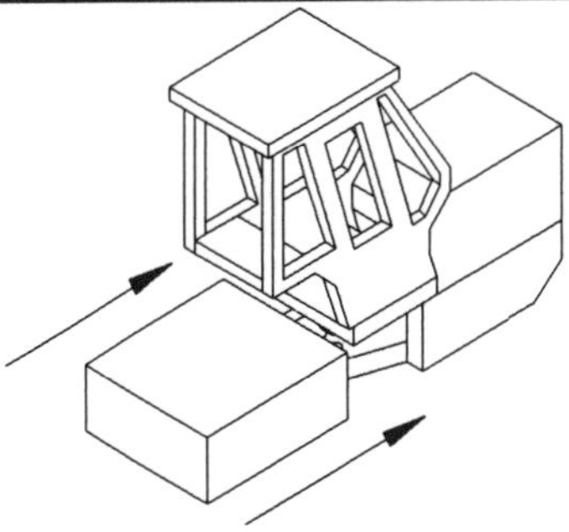

Slide front assembly into rear body.

12

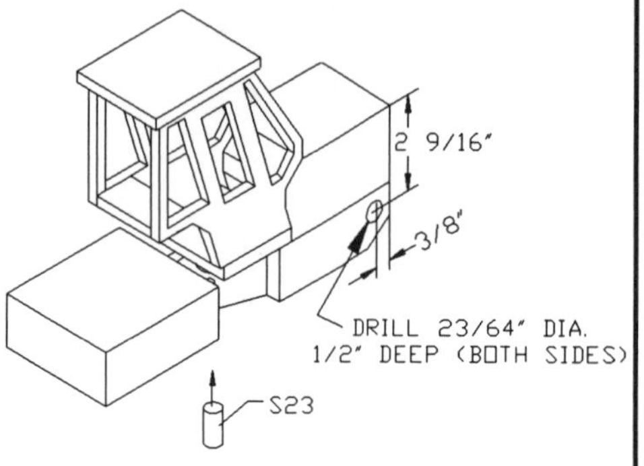

Insert main pivot pin S23. Drill holes into body, as shown.

13

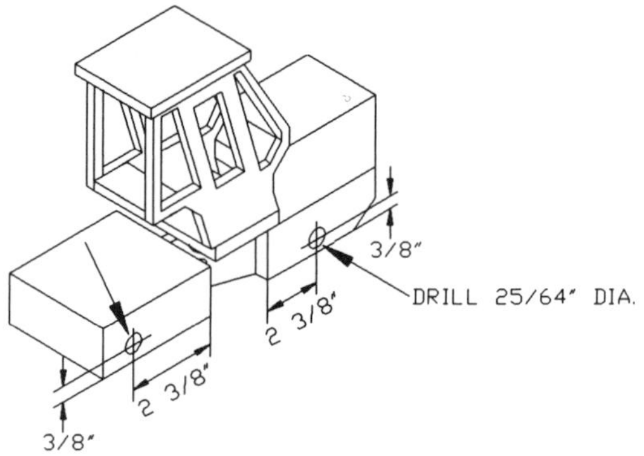

Drill holes into body, as shown.

14

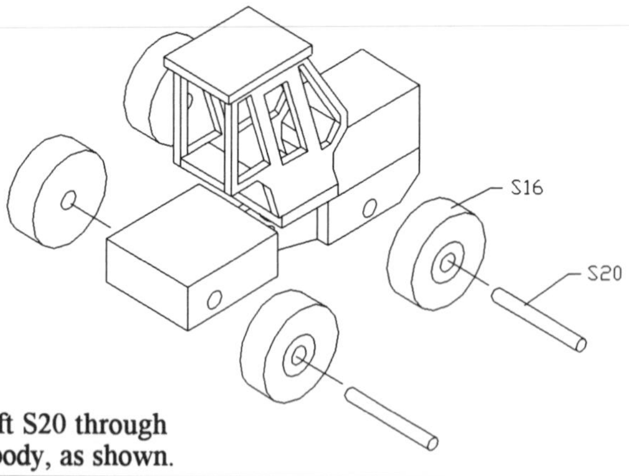

Insert wheel shaft S20 through wheels S16 and body, as shown.

15

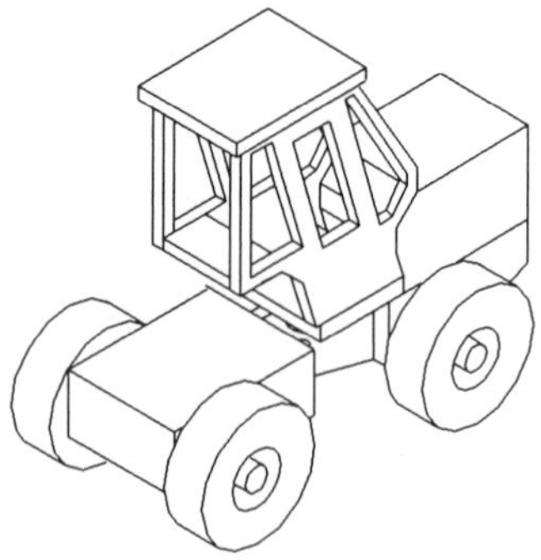

Start shovel.

16

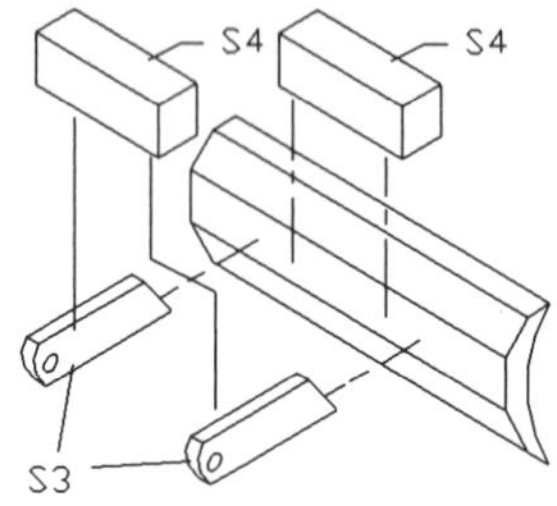

Glue levers S3 and re-enforcing blocks S4.

17

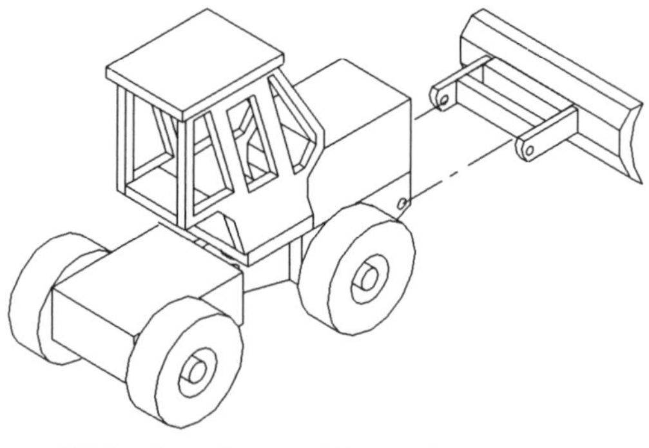

Slide shovel assembly, as shown.

18

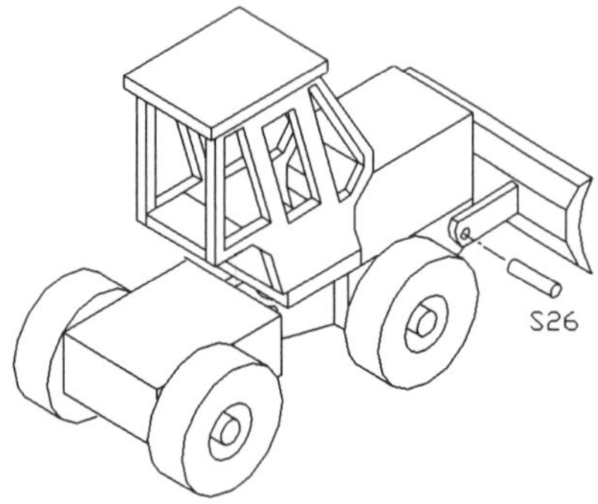

Insert pins S26.

19

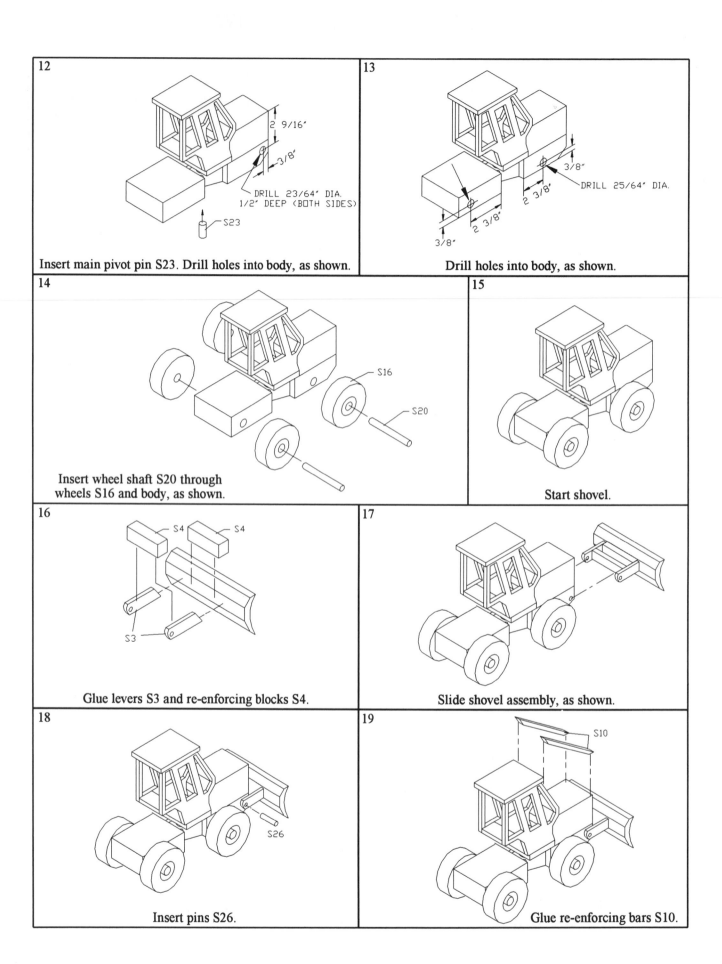

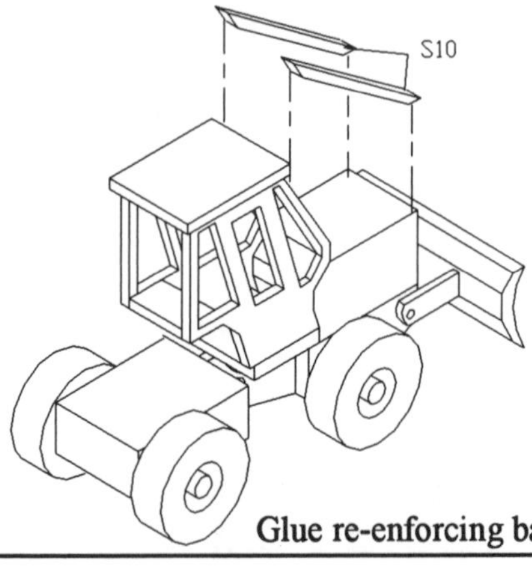

Glue re-enforcing bars S10.

20

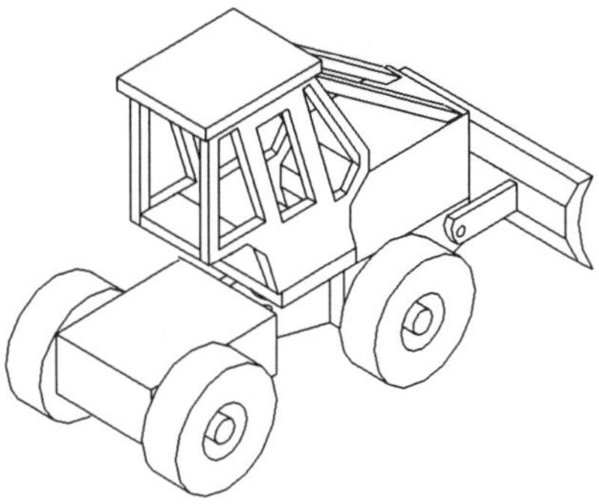

Start pulley assembly.

21

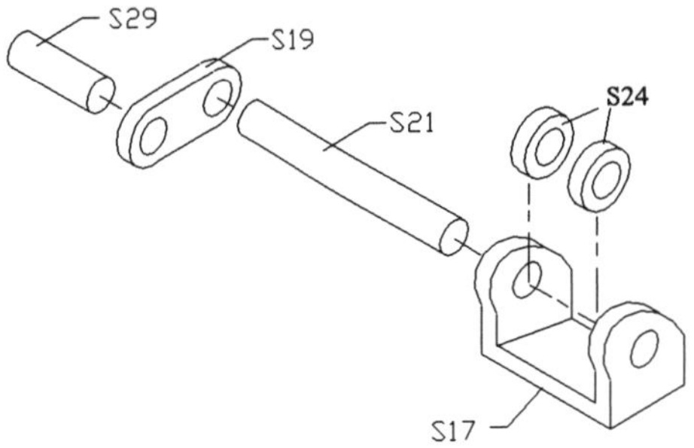

Assemble rear pulley assembly, as shown.

22

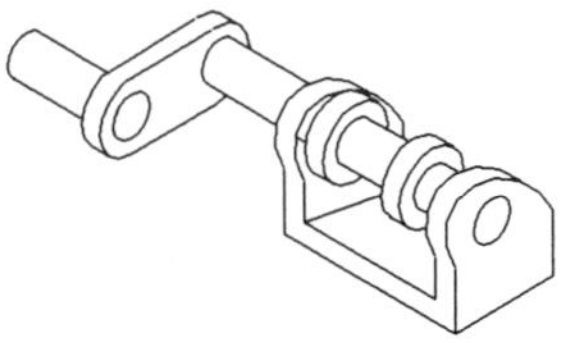

Rear pulley shown assembled.

23

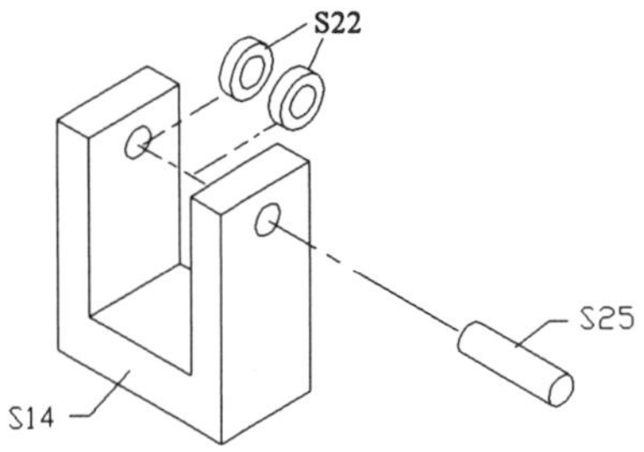

Assemble pulley parts, as shown.

24

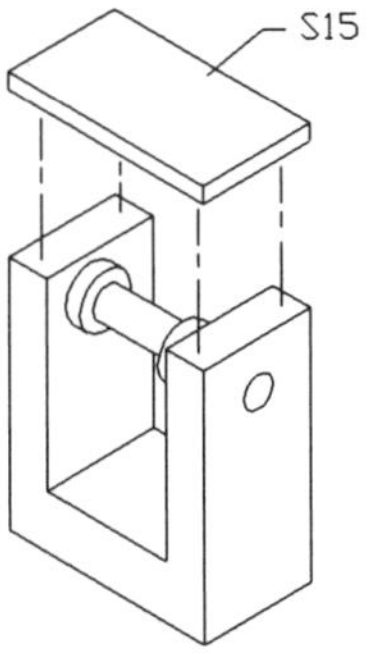

Glue pulley plate S15.

25

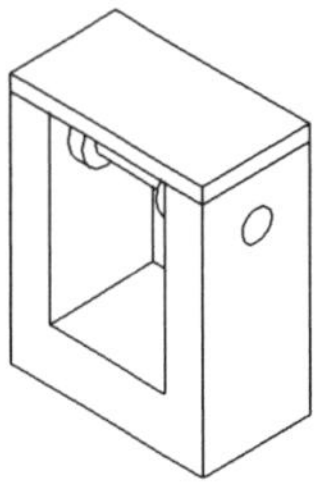

Front pulley assembly shown assembled.

26

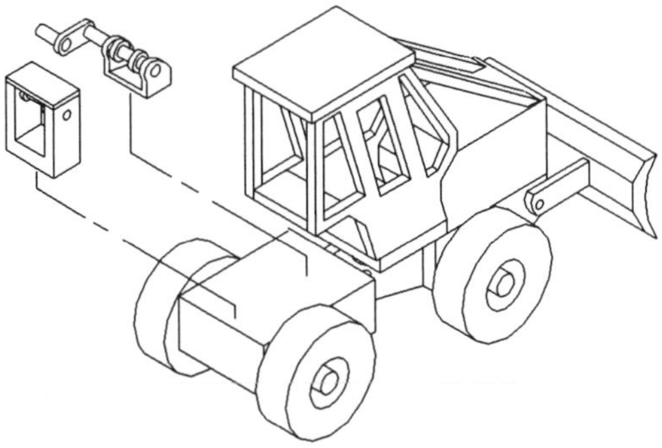

Glue parts, as shown.

27

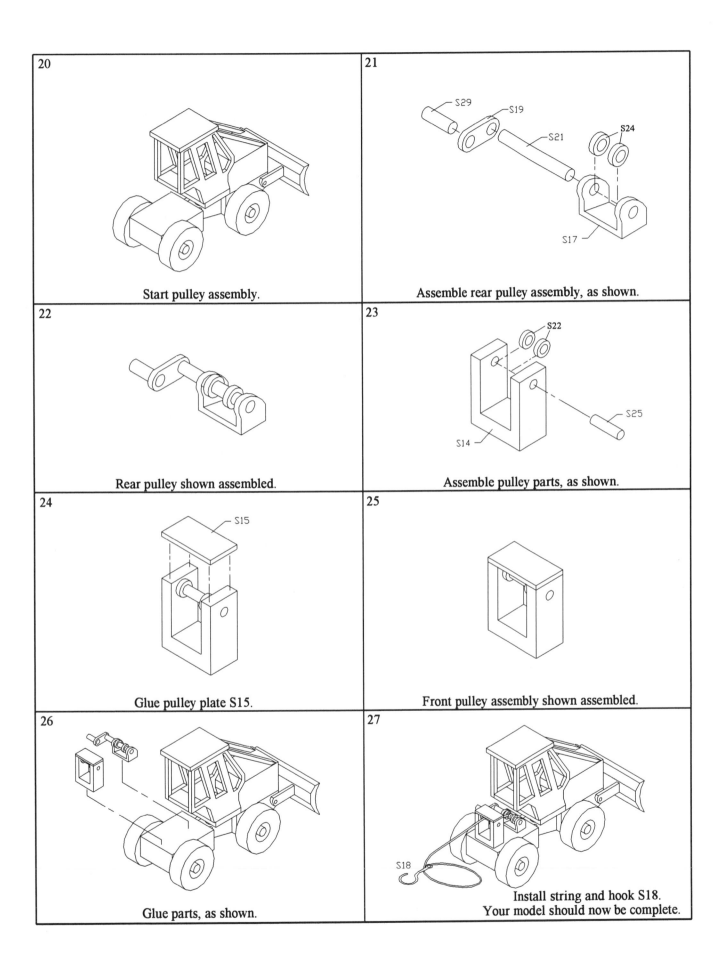

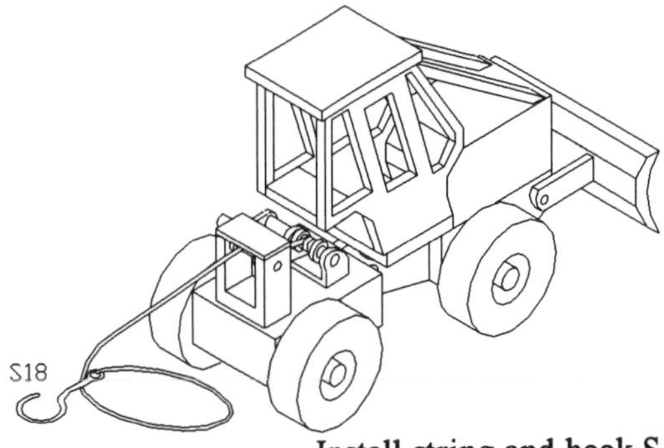

Install string and hook S18.
Your model should now be complete.

GRAPPLE SKIDDER

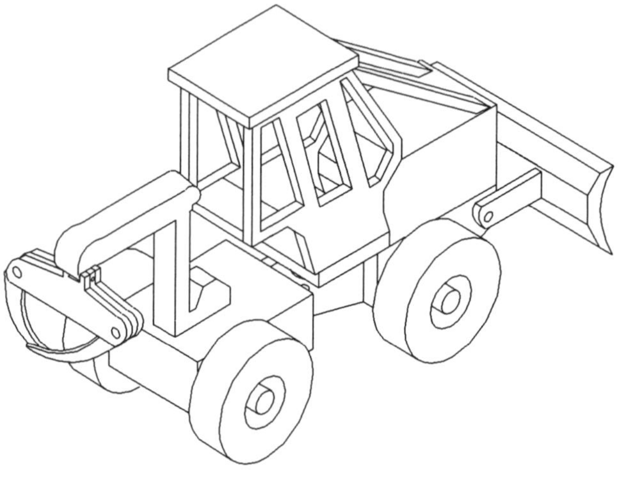

General Instructions - Grapple Skidder

1- Start by cutting materials needed by following the list of materials, paying attention to the rough and finished size. **Identify the parts as they are cut.**

Please note: Different types of wood can be used for the various parts. It is suggested, however, that hard wood be used, since many of the parts would be much too fragile if using soft wood. We have used a combination of pine, maple and oak to give the models a nice contrast!

2- Remove the full-size patterns found in the appendix. Cut them out, leaving approximately 1/16" all around, and place on the proper piece of wood. Patterns can be secured to wood using either spray adhesive or rubber ciment. If using the latter, cut and sand the part first to finished size. If drilling is required, mark the hole by inserting a scriber or nail through the pattern into the wood. Remove the pattern before drilling.

You should have no trouble determining which surface to attach most of the patterns. Some parts, however, can be confusing since the pattern could fit on more than one surface. The drawings below indicate exactly which surface to attach the patterns for these parts.

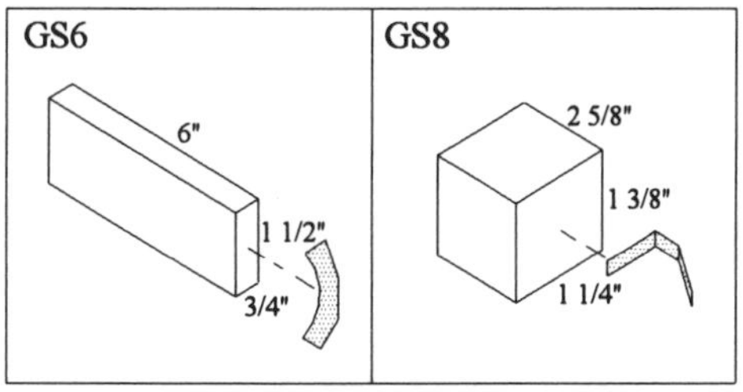

3- Look at the full-size drawing sheets to finish parts GS2 and GS26.

4- Using maple dowels, make all pins, shafts, etc.

5- Follow the assembly drawings to complete your model.

List of Materials - Grapple Skidder

Part	T	W	L	Material	Qty.	*
GS1	3/8"	3"	6 7/8"	pine	2	R
GS2	3/4"	2 5/8"	5 1/2"	pine	1	F
GS3	3/8"	1"	3"	maple	2	R
GS4	1/2"	5/8"	3 3/8"	maple	2	F
GS5	3/8"	1 1/2"	5 1/4"	pine	2	F
GS6	3/4"	1 1/2"	6"	pine	1	F
GS7	1 1/2"	3 3/8"	3 3/4"	pine	1	F
GS8	1 1/4"	1 3/8"	2 5/8"	pine	1	F
GS9	3/4"	1 5/8"	2 3/4"	maple	1	R
GS10	1/4"	3 3/8"	2 5/8"	pine	1	F
GS11	3/8"	1/2"	4 5/8"	oak	2	R

Part	T	W	L	Material	Qty.	*
GS12	1 3/4"	3 3/8"	3 15/16"	pine	1	F
GS13	3/8"	3 3/4"	3 1/8"	pine	1	F
GS14	3/8"	4"	4"	oak	2	R
GS15	1/4"	2 3/4"	4 1/4"	maple	2	R
GS16	1/4"	1 3/4"	1 3/4"	maple	1	R
GS17	1/4"	1 3/4"	1 3/4"	maple	1	R
GS18	1/2"	2 3/4"	4 1/4"	maple	1	R
GS20	1/4"	1 7/8"	4 7/8"	maple	2	R
GS21	3/4"	4"	5 3/8"	maple	1	R
GS26	1 1/4"	3 3/4 DIA.		oak	4	R

R = Rough size
F = Finished size

T = Thickness
W = Width
L = Length

Instructions:

R= Rough sizes, the material is cut oversized so you have ample room to apply the pattern on the surface. Sanding is not required at this point.

F = Finished Size: Cut and sand parts to finished size.

Full-Sized Patterns: Set One

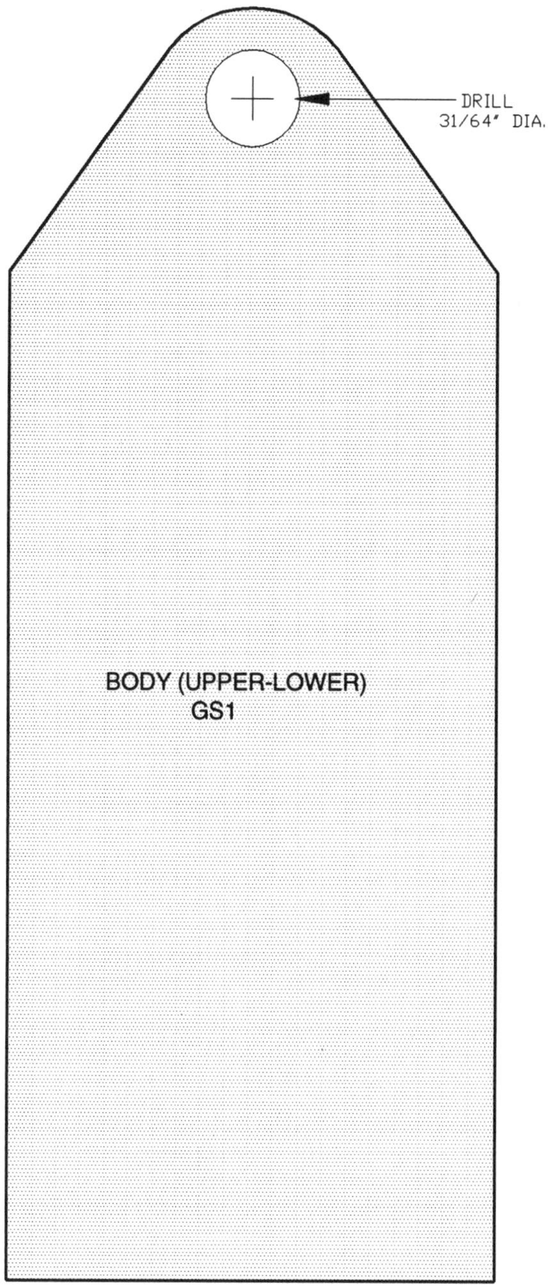

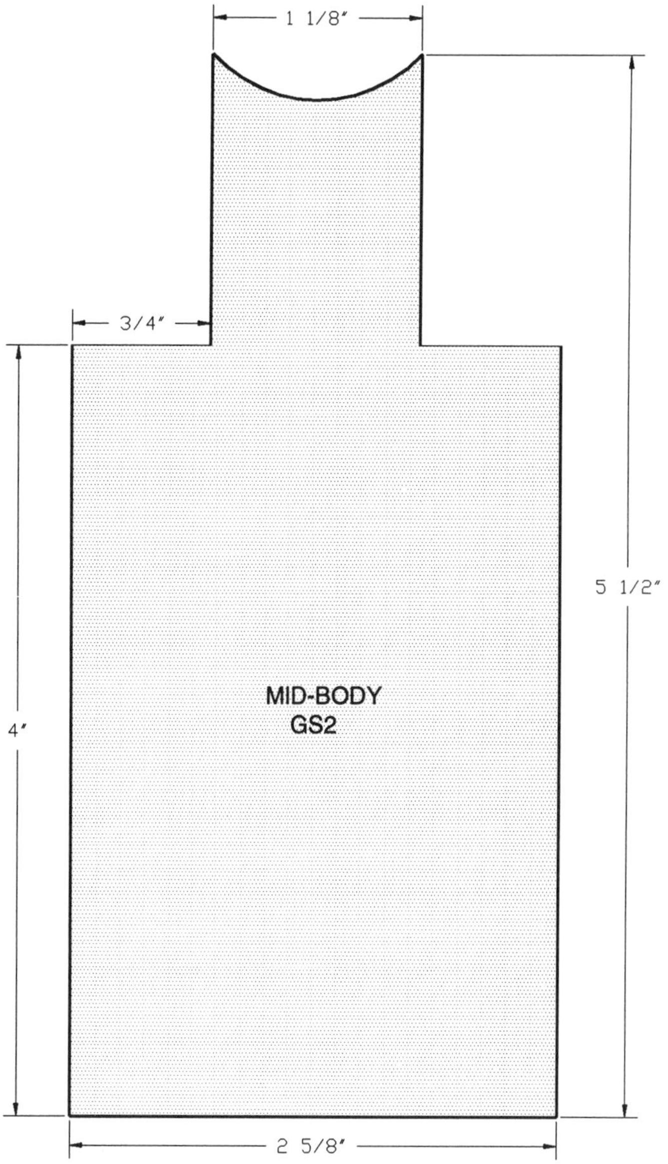

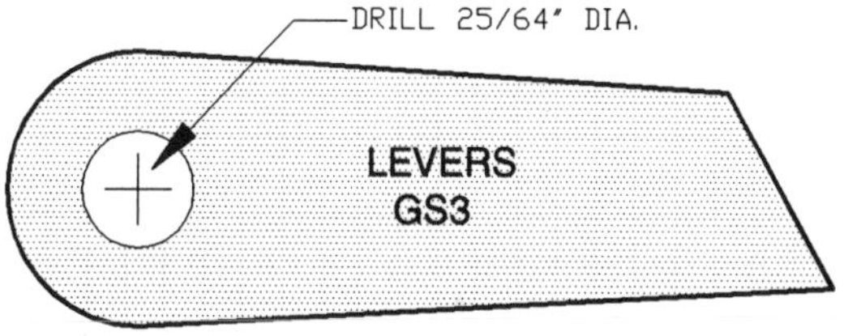

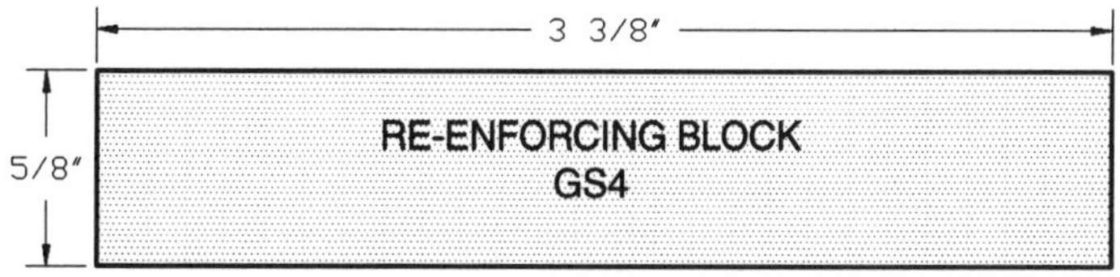

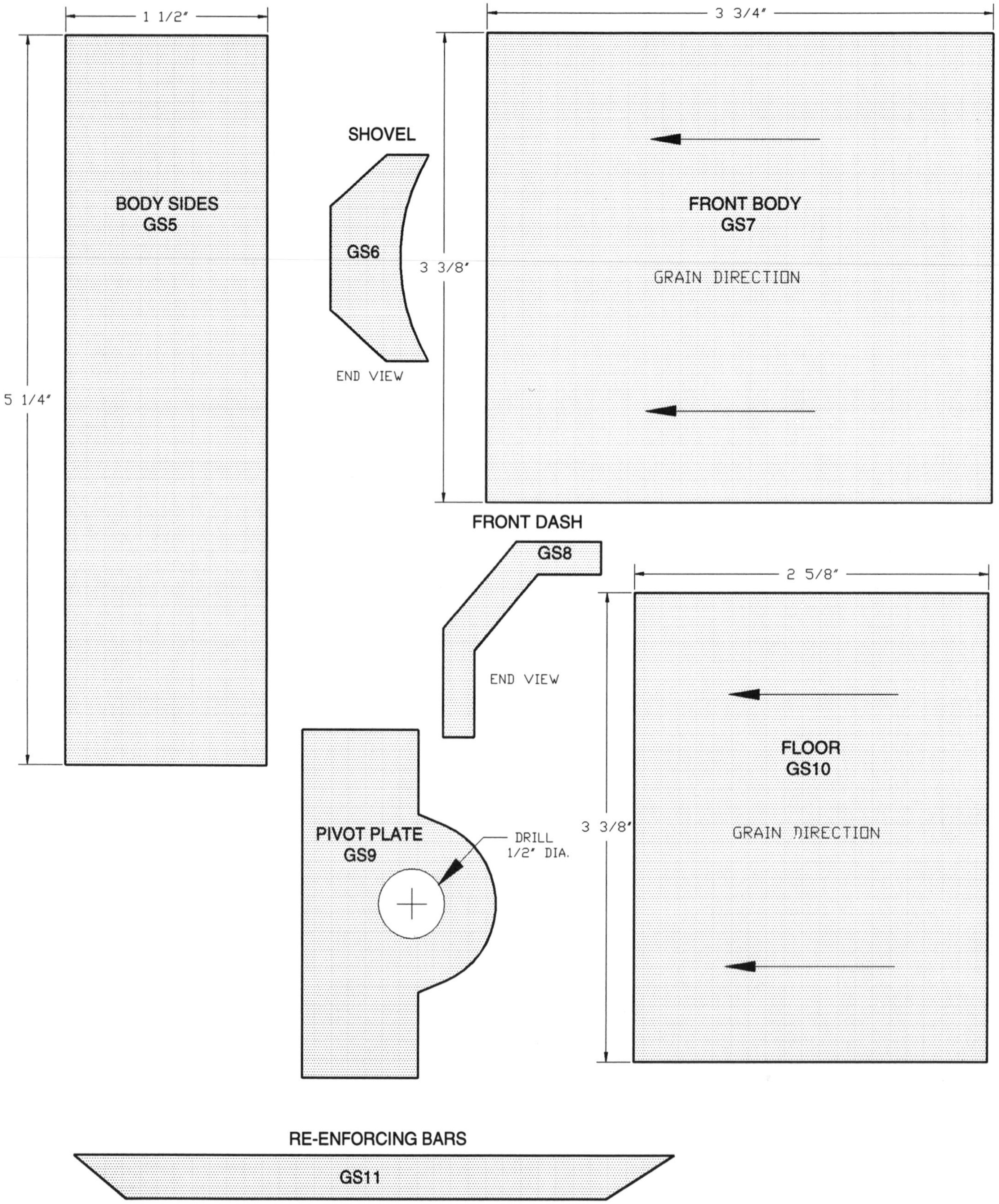
1 1/2"
5 1/4"
BODY SIDES
GS5
SHOVEL
GS6
END VIEW
3 3/4"
3 3/8"
FRONT BODY
GS7
GRAIN DIRECTION
FRONT DASH
GS8
END VIEW
2 5/8"
3 3/8"
FLOOR
GS10
GRAIN DIRECTION
PIVOT PLATE
GS9
DRILL
1/2" DIA.
RE-ENFORCING BARS
GS11

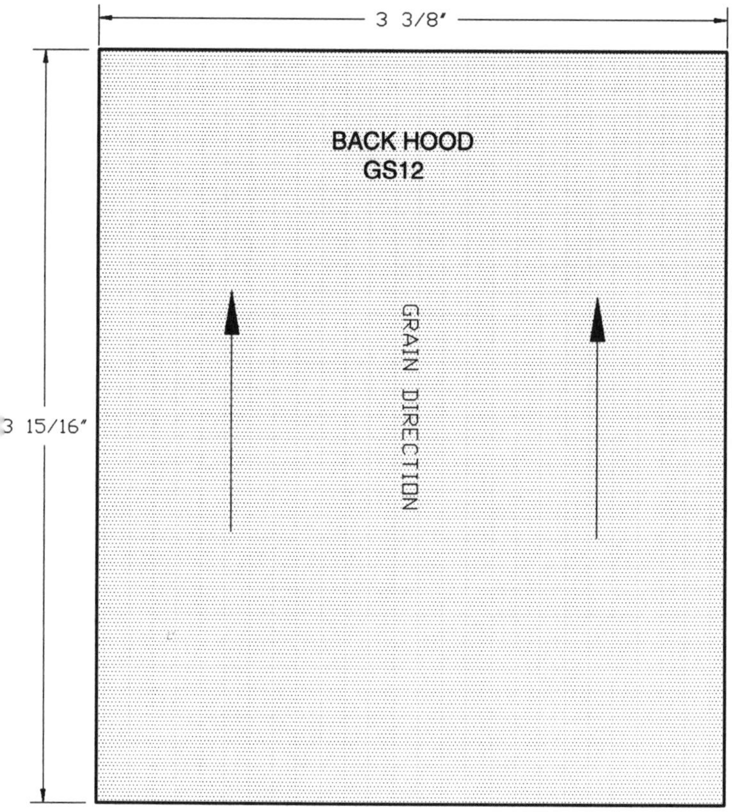
3 3/8"
3 15/16"
BACK HOOD
GS12
GRAIN DIRECTION

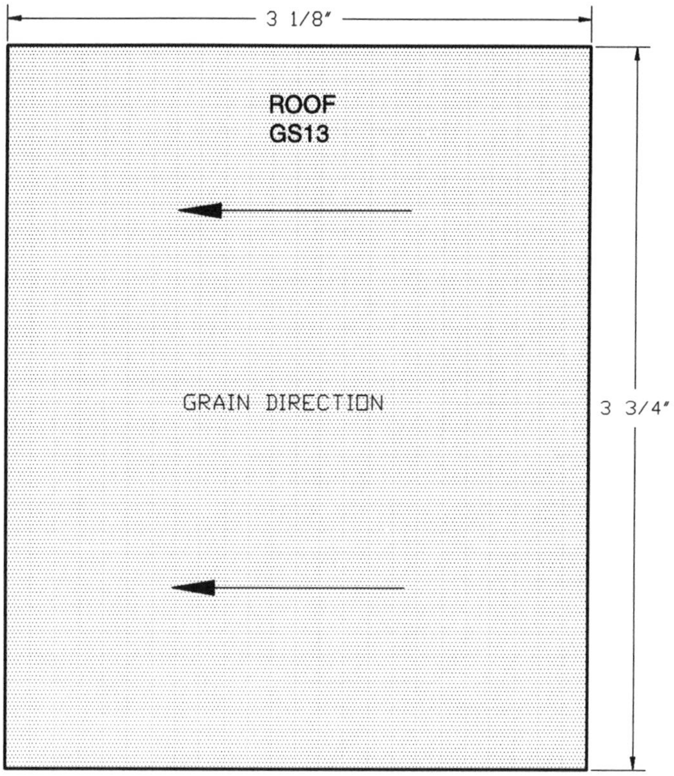
3 1/8"
ROOF
GS13
GRAIN DIRECTION
3 3/4"

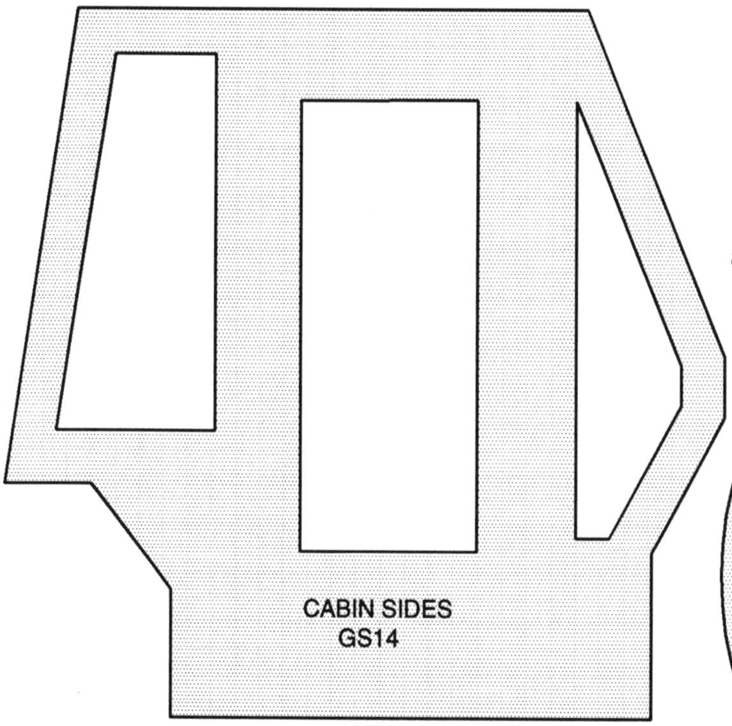
CABIN SIDES
GS14

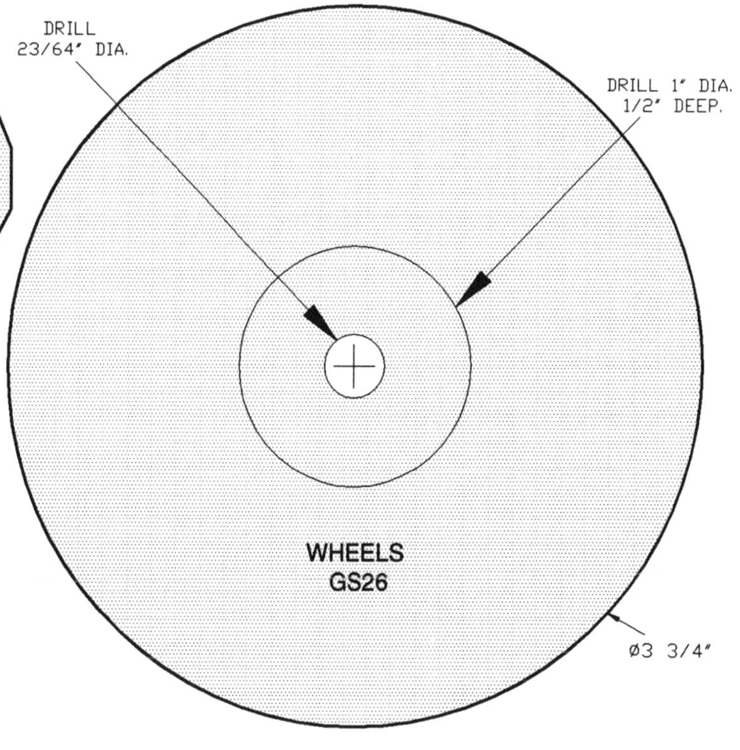
DRILL
23/64" DIA.
DRILL 1" DIA.
1/2" DEEP.
WHEELS
GS26
Ø3 3/4"

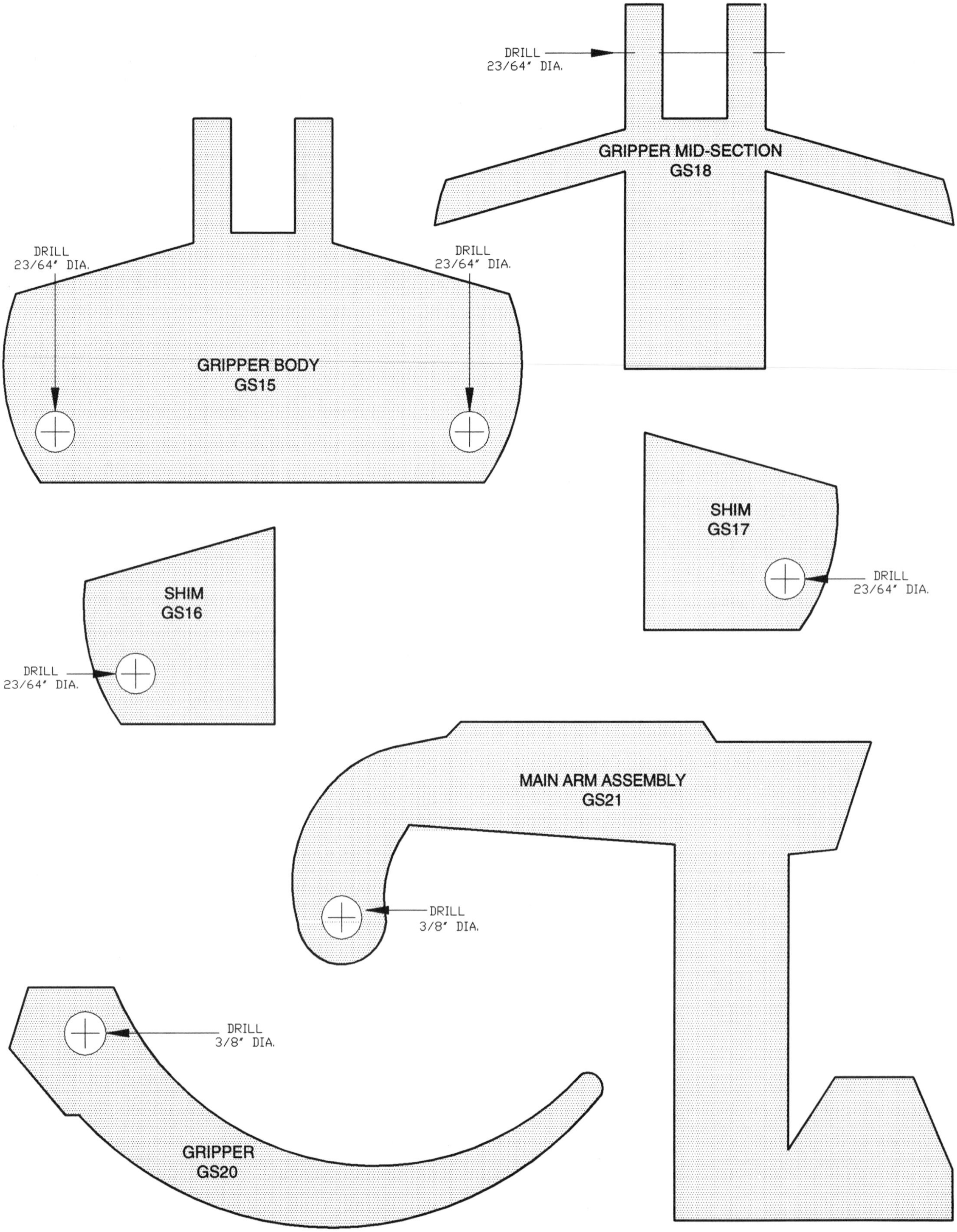
DRILL
23/64" DIA.
GRIPPER MID-SECTION
GS18
DRILL
23/64" DIA.
DRILL
23/64" DIA.
GRIPPER BODY
GS15
SHIM
GS17
DRILL
23/64" DIA.
SHIM
GS16
DRILL
23/64" DIA.
MAIN ARM ASSEMBLY
GS21
DRILL
3/8" DIA.
DRILL
3/8" DIA.
GRIPPER
GS20

GS27

WHEEL SHAFTS
(2 REQUIRED)

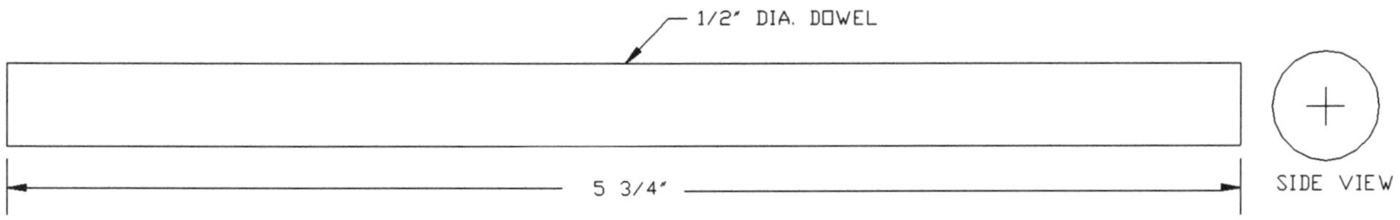

GS28

MAIN PIVOT PIN
(1 REQUIRED)

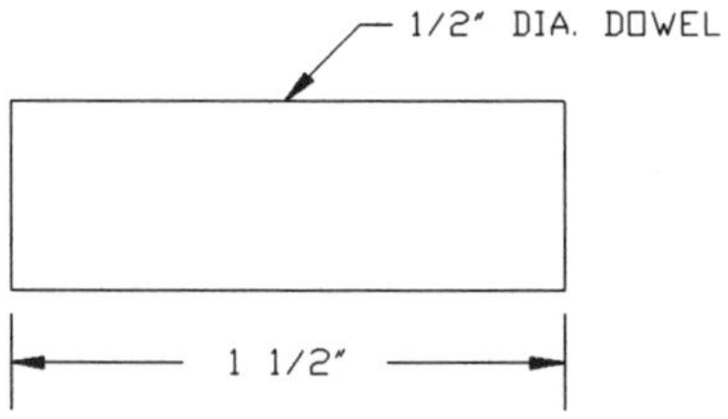

GS29

PINS
(2 REQUIRED)

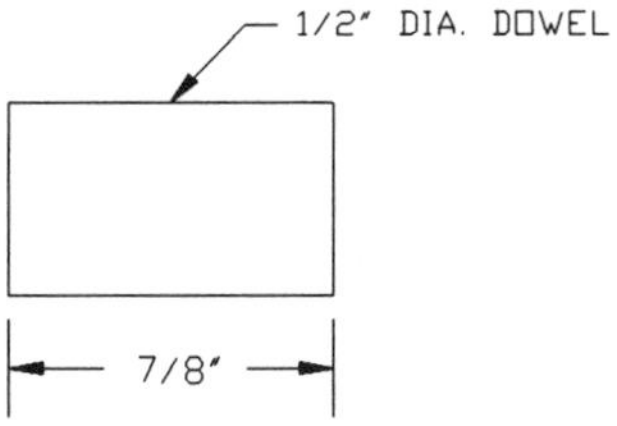

GS33

PIN
(1 REQUIRED)

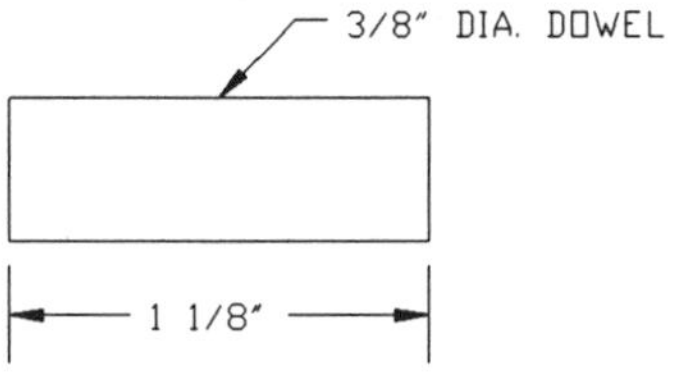

GS34

PINS
(2 REQUIRED)

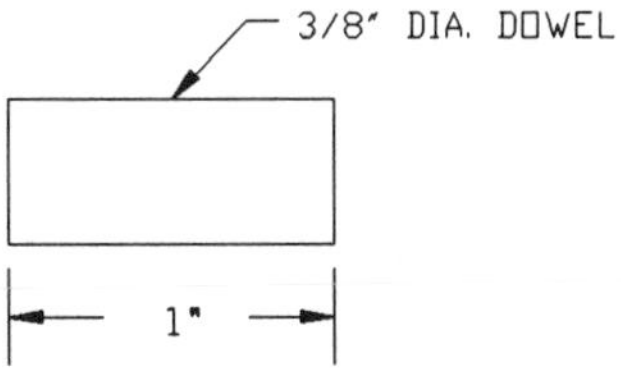

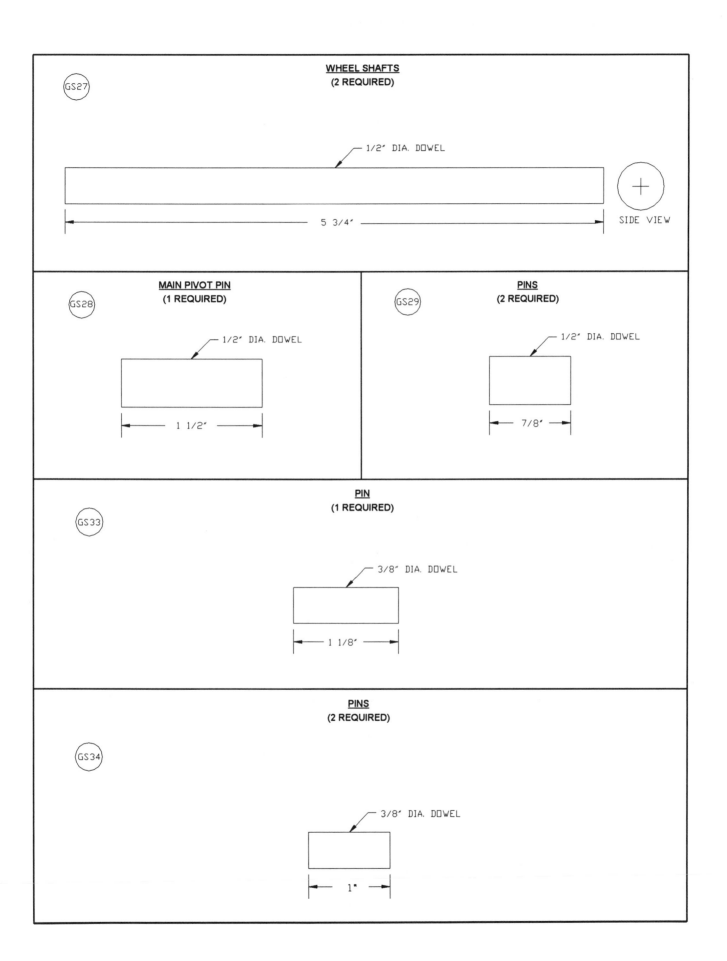

Grapple Skidder - Assembly Drawings

1

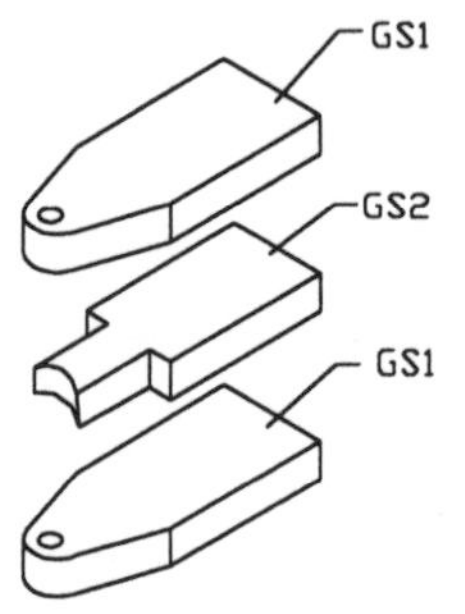

Glue mid-body GS2 between
upper and lower body GS1.

2

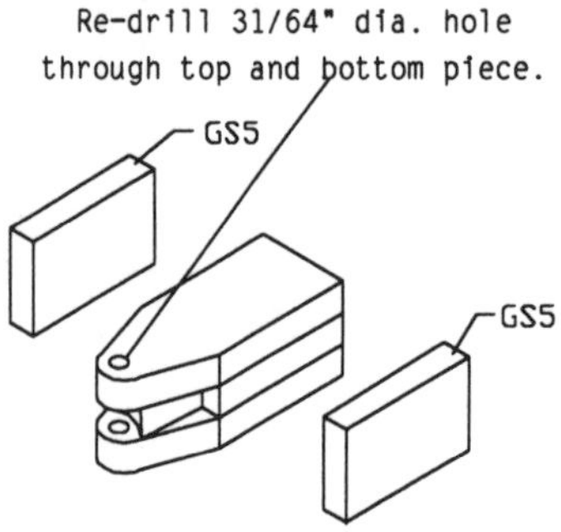

Re-drill hole, as shown.
Glue sides GS5.

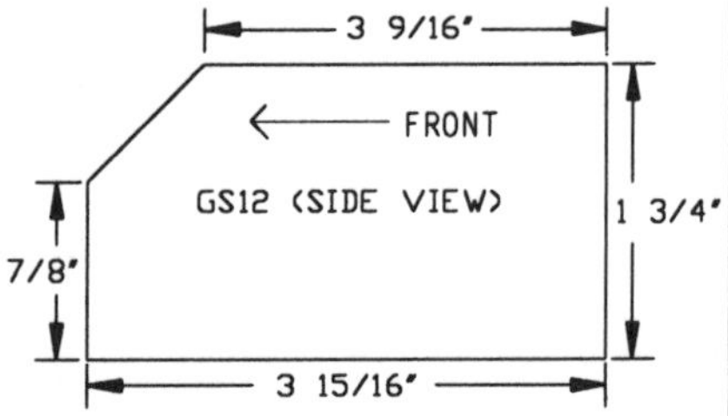

Sand angle on rear hood, as shown.

3

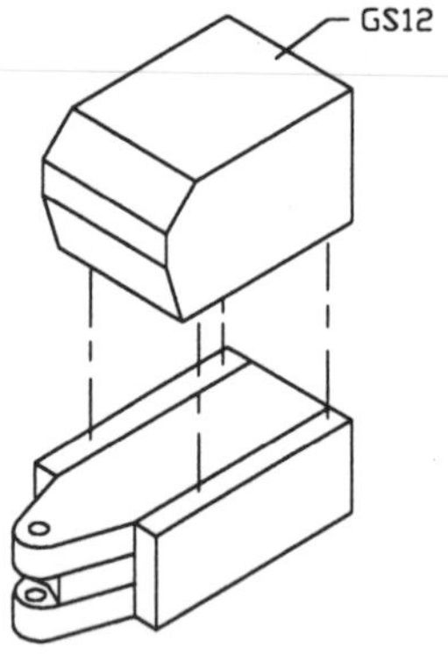

Glue rear hood GS12 onto body.

4

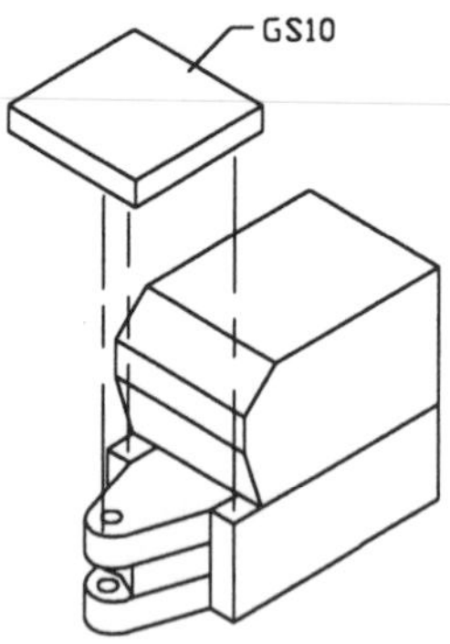

Glue cabin floor GS10.

5

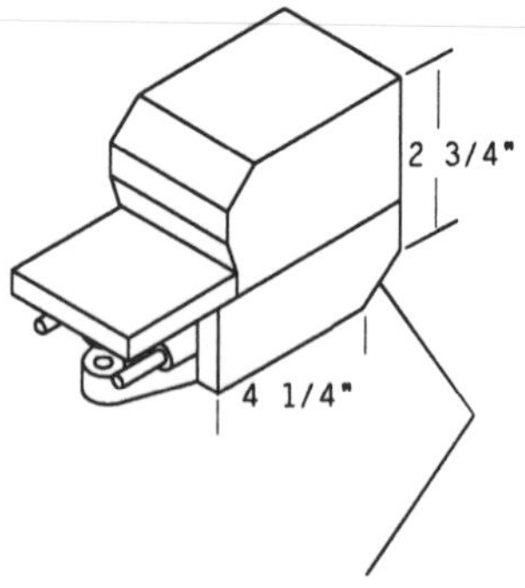

Sand angle (from side to side).

6

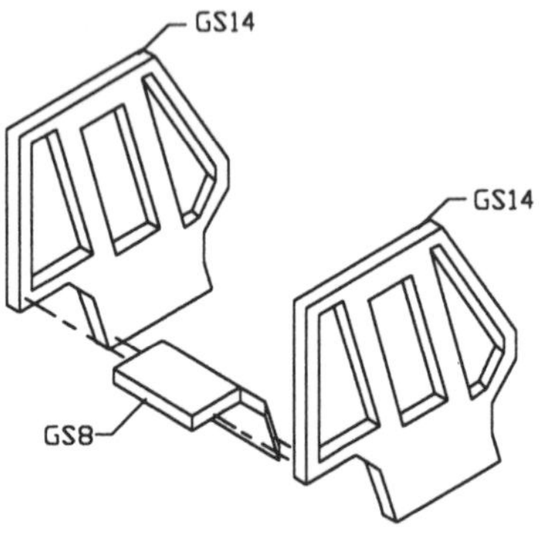

Glue front dash GS8, as shown.

7

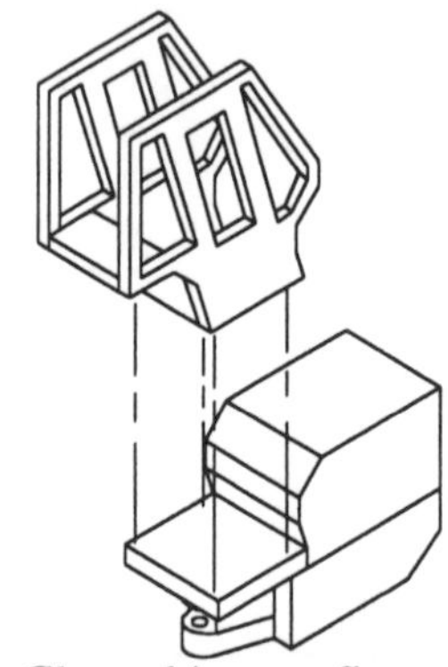

Glue cabin onto floor.

8

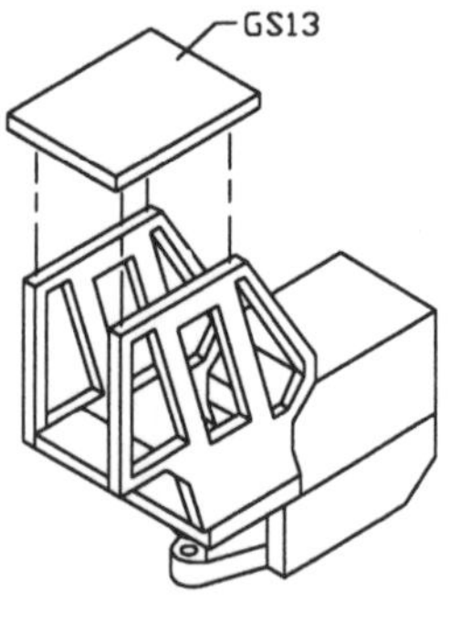

Glue roof GS13 onto cabin.

9

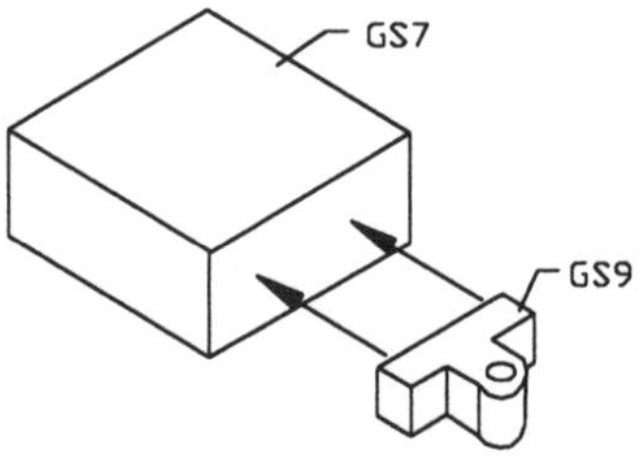

Glue pivot plate GS9 onto body GS7.
See next frame for exact location.

10

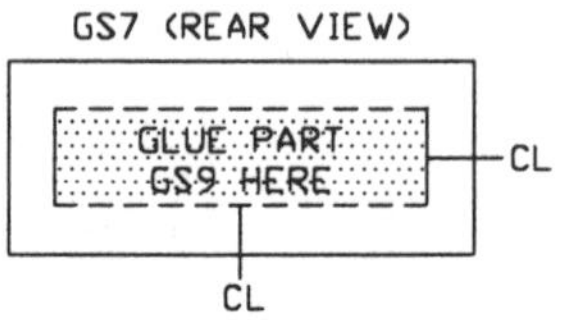

Where to glue pivot
plate GS9 onto body GS7.

11

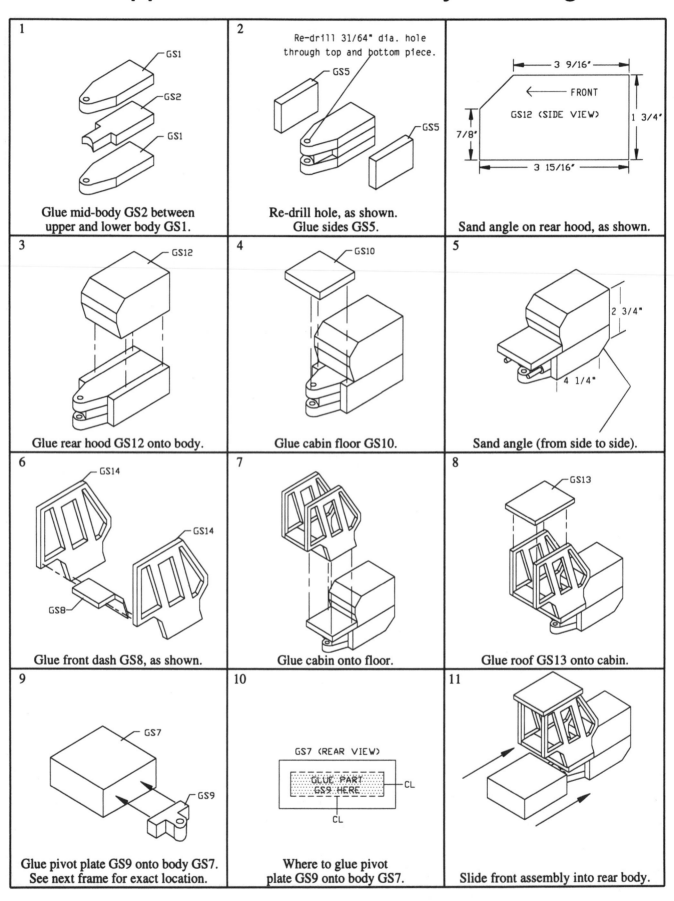

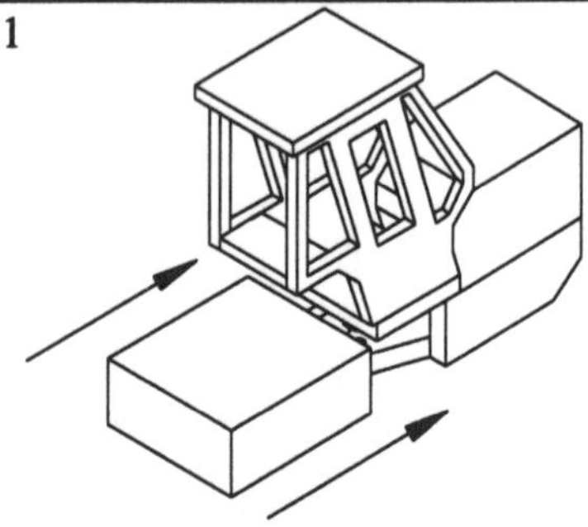

Slide front assembly into rear body.

12

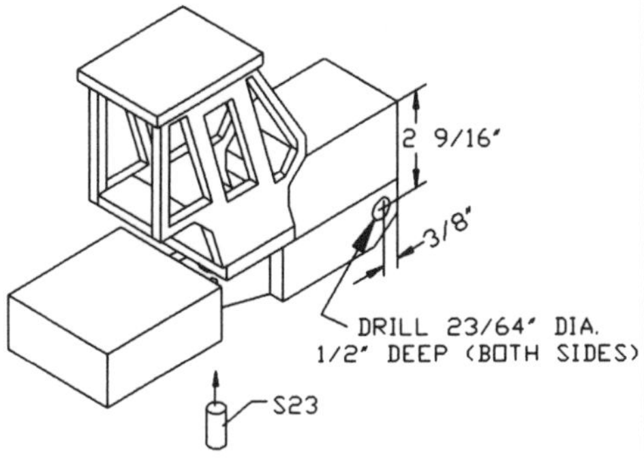

Insert main pivot pin GS28. Drill holes into body, as shown.

13

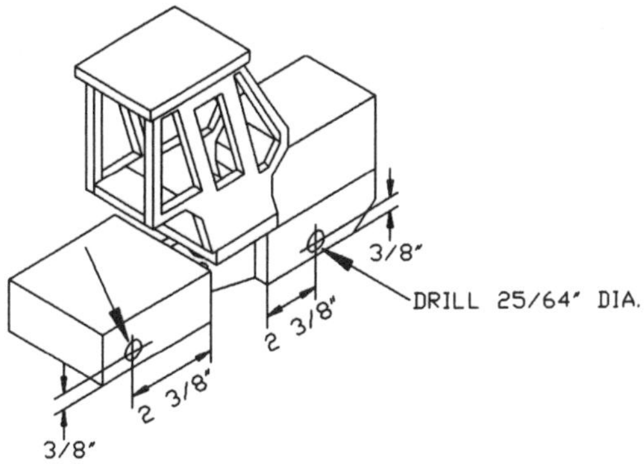

Drill holes into body, as shown.

14

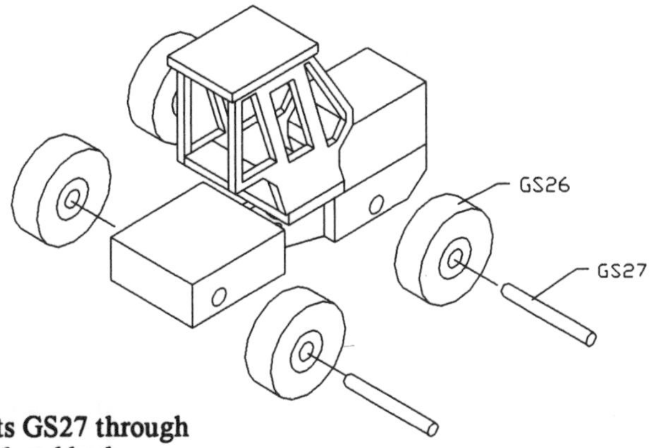

Insert wheel shafts GS27 through wheels GS26 and body.

15

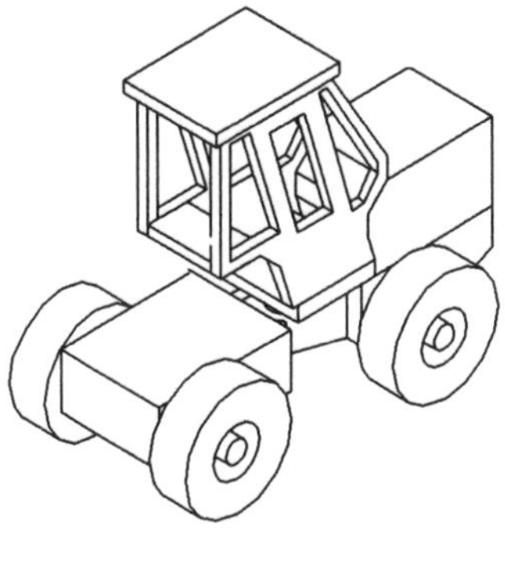

Start shovel assembly.

16

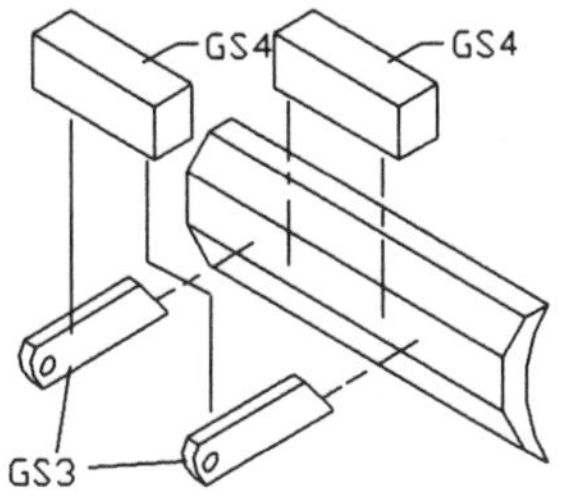

Glue levers GS3 and re-enforcing blocks GS4, as shown.

17

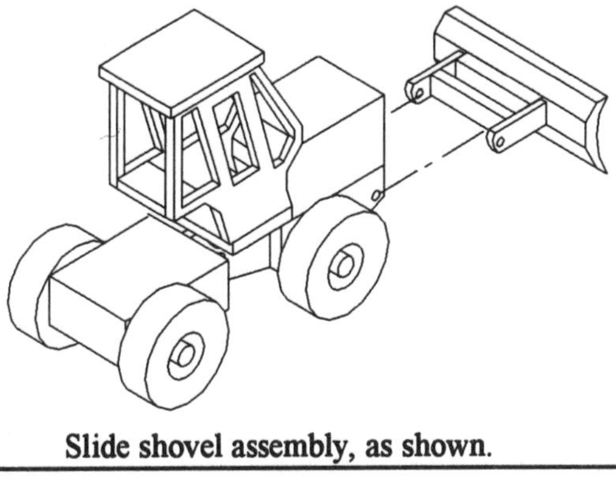

Slide shovel assembly, as shown.

18

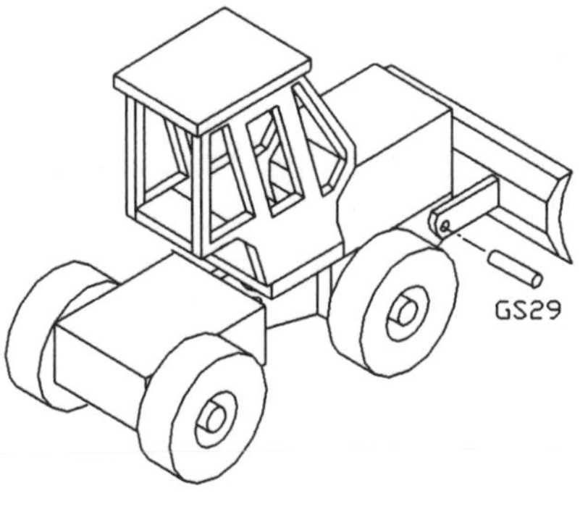

Insert pins GS29.

19

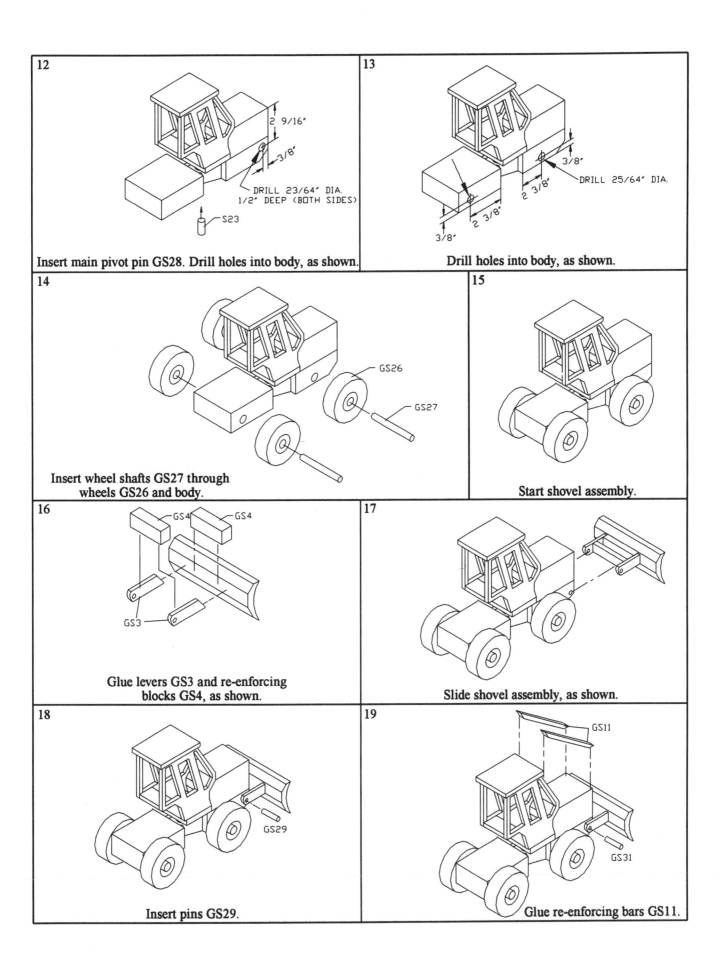

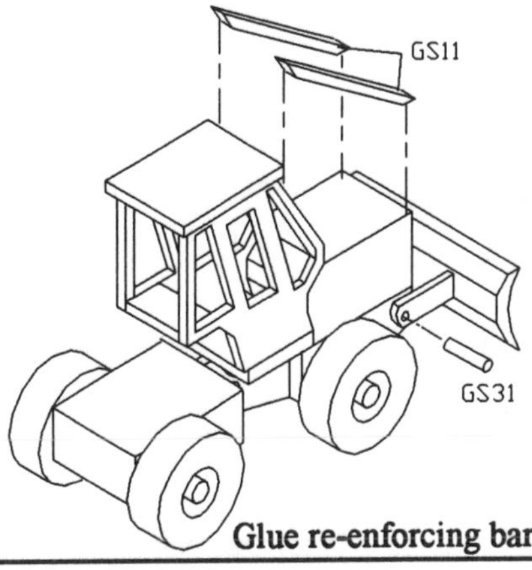

Glue re-enforcing bars GS11.

20

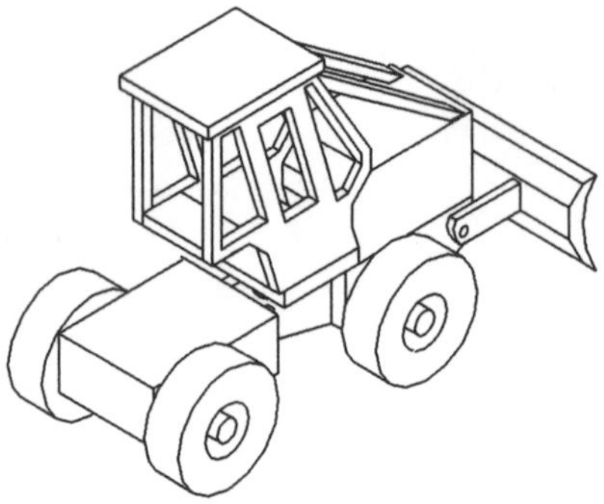

Start gripper assembly.

21

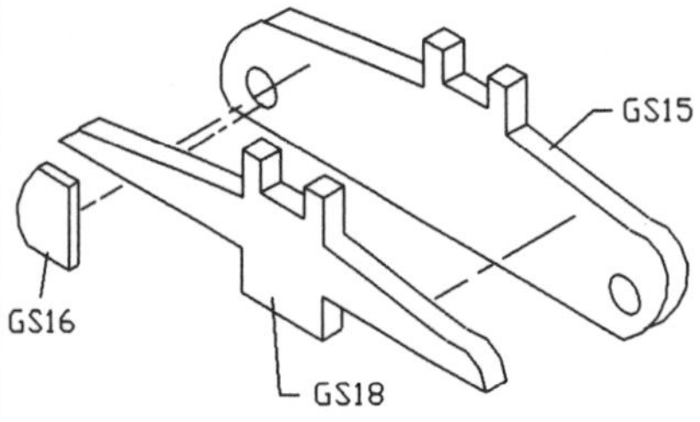

Glue shim GS16, mid-section GS18 onto gripper body GS15.

22

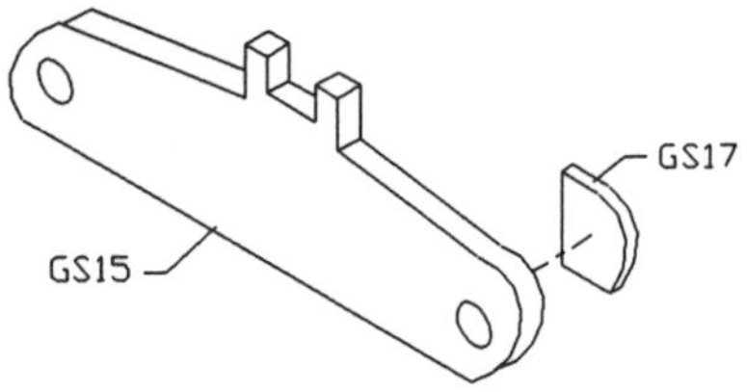

Glue other shim GS17 onto other gripper body GS15.

23

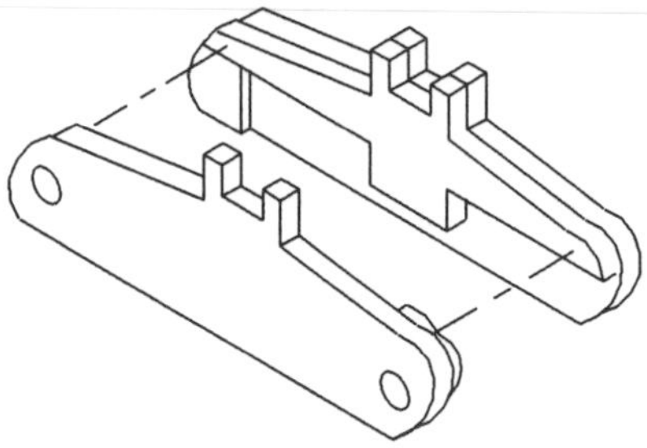

Glue both assembly's together.

24

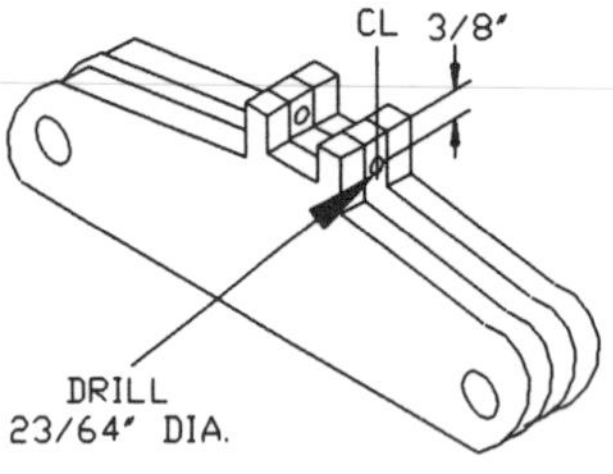

Drill hole, as shown.

25

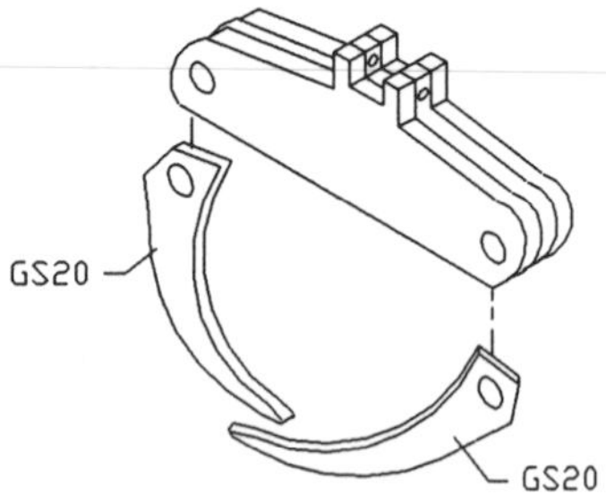

Slide grippers GS20 into gripper body assembly.

26

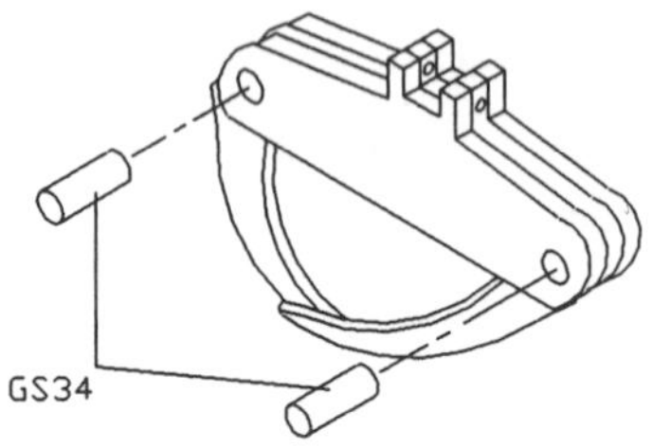

Insert pins GS34.

27

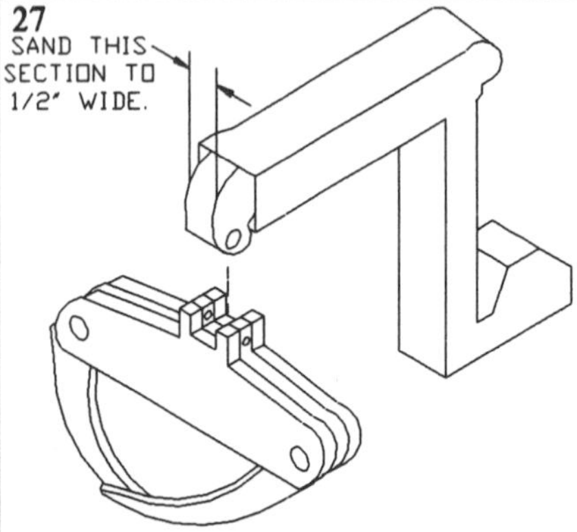

Sand part, as shown. Slide in assembly.

28

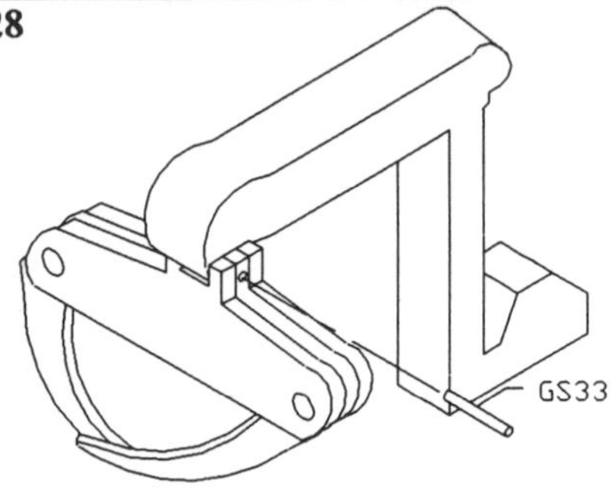

Insert pins GS33.

29

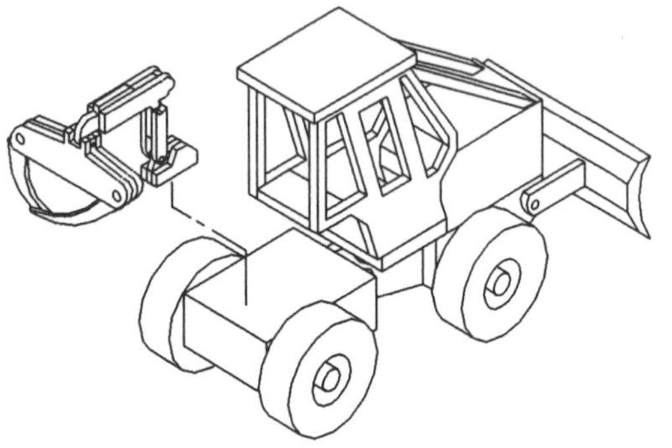

Glue gripper assembly onto body.

30

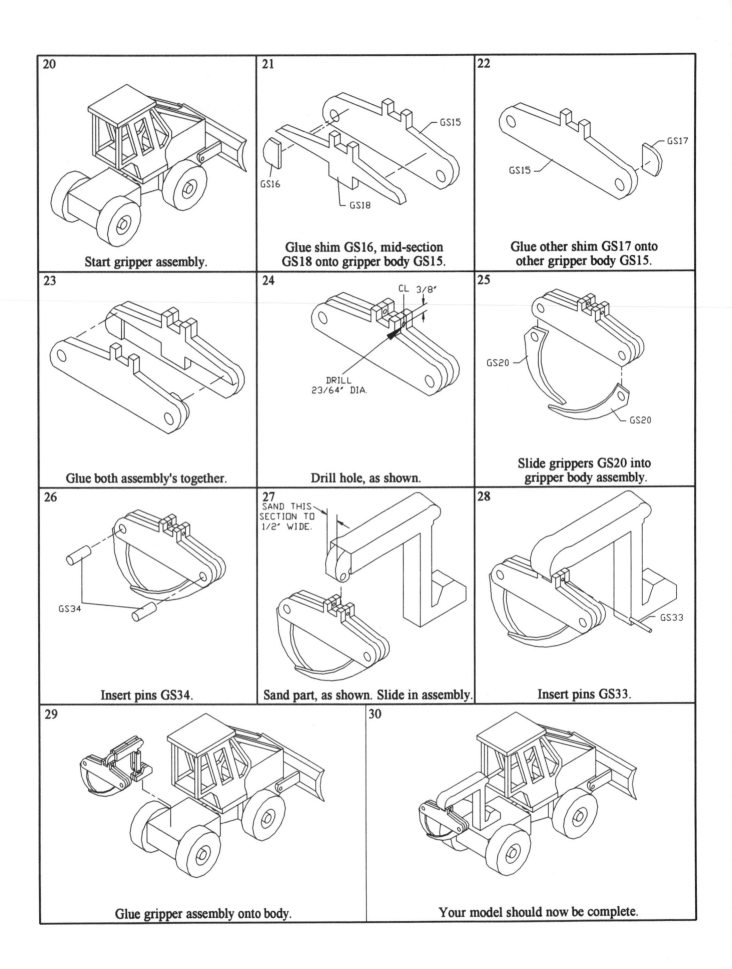

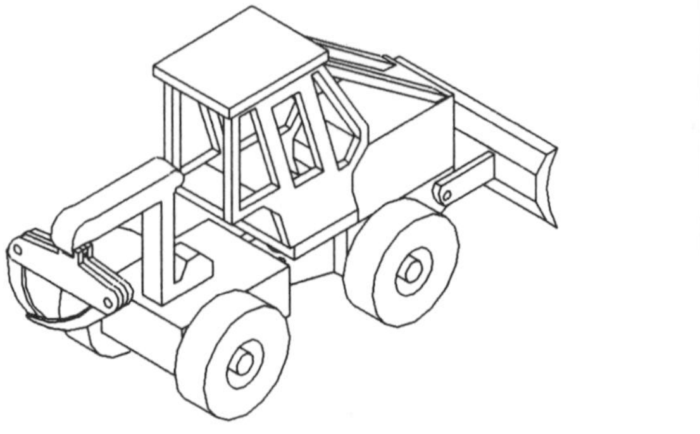

Your model should now be complete.

BACKHOE

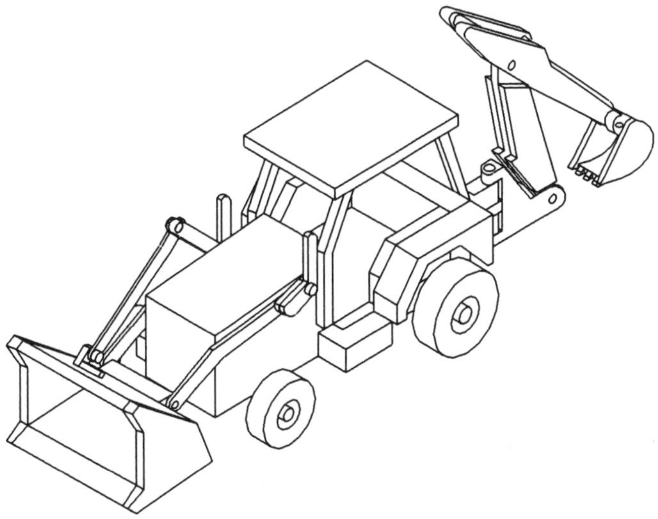

General Instructions - Backhoe

1- Start by cutting materials needed by following the list of materials, paying attention to the rough and finished size. **Identify the parts as they are cut.**

Please note: Different types of wood can be used for the various parts. It is suggested, however, that hard wood be used, since many of the parts would be much too fragile if using soft wood. We have used a combination of pine, maple and oak to give the models a nice contrast!

2- Remove the full-size patterns found in the appendix. Cut them out, leaving approximately 1/16" all around, and place on the proper piece of wood. Patterns can be secured to wood using either spray adhesive or rubber ciment. If using the latter, cut and sand the part first to finished size. If drilling is required, mark the hole by inserting a scriber or nail through the pattern into the wood. Remove the pattern before drilling.

You should have no trouble determining which surface to attach most of the patterns. Some parts, however, can be confusing since the pattern could fit on more than one surface. The drawings below indicate exactly which surface to attach the patterns for these parts.

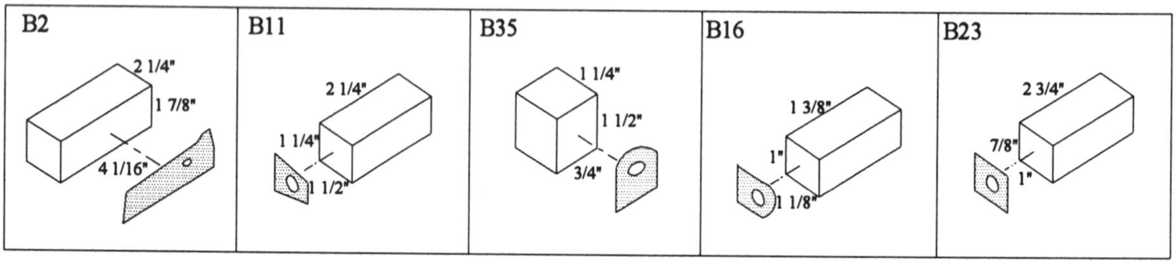

3- Look at the full-size drawing sheets to finish part B3.

4- Parts B11, B35, B14, B16, B19 and B23 will need additional cuts and details, please refer to the additional information pages, to complete these parts.

5- Using maple dowels, make all pins, shafts, etc.

6- Follow the assembly drawings to complete your model.

List of Materials - Backhoe

Part	T	W	L	Material	Qty.	*
B1	1"	2 1/4"	7 1/2"	pine	1	F
B2	1 7/8"	2 1/4"	4 1/16"	pine	1	F
B3	5/8"	2 3/4"	2 3/8"	pine	1	F
B4	1/4"	1"	3 3/8"	pine	2	F
B5	1"	1 1/2"	3 5/8"	pine	2	R
B6	3/8"	4"	5"	oak	2	R
B7	3/8"	1 1/4"	2 3/4"	maple	1	R
B8	3/8"	3 1/2"	4 1/4"	pine	1	F
B9	1/4"	2"	5 1/2"	pine	1	F
B10	1/4"	1 1/2"	6 1/8"	pine	1	F
B11	1 1/4"	1 1/2"	2 1/4"	maple	1	F
B12	1/4"	2 5/8"	5 1/2"	pine	2	F
B13	1/4"	2 5/8"	2 3/4"	pine	1	R
B14	1"	2 1/2"	2 1/2"	maple	1	R
B15	1"	1 1/2"	1 3/8"	oak	2	R
B16	7/8"	1 1/4"	1 3/8"	maple	1	F

Part	T	W	L	Material	Qty.	*
B17	3/8"	1 3/4"	6 5/8"	maple	2	R
B18	1/4"	2"	3 1/8"	maple	2	R
B19	1 3/4"	2 5/8"	3 1/2"	maple	1	R
B20	1/4"	1 3/4"	4 1/2"	maple	2	R
B21	3/8"	1 7/8"	5 1/2"	maple	1	R
B23	7/8"	1"	2 3/4"	maple	1	F
B24	1/4"	1 1/2"	3"	oak	2	R
B25	1 1/4"	3 1/8" DIA.		oak	2	F
B26	1"	1 7/8" DIA.		oak	2	F
B27	1/4"	1 1/2"	4 5/8"	maple	2	R
B28	3/8"	1 1/2"	6 1/4"	maple	1	R
B33	1/4"	1"	6 3/4"	maple	1	R
B34	3/8"	2 3/4"	2 3/4"	maple	1	R
B35	1"	1 1/4"	1 3/8"	maple	1	F
B36	3/8"	3"	3 1/4"	maple	1	R

T = Thickness
W = Width
L = Length

R = Rough size
F = Finished size

Instructions:

R= Rough sizes, the material is cut oversized so you have ample room to apply the pattern on the surface. Sanding is not required at this point.

F = Finished Size: Cut and sand parts to finished size.

Full-Sized Patterns: Set One

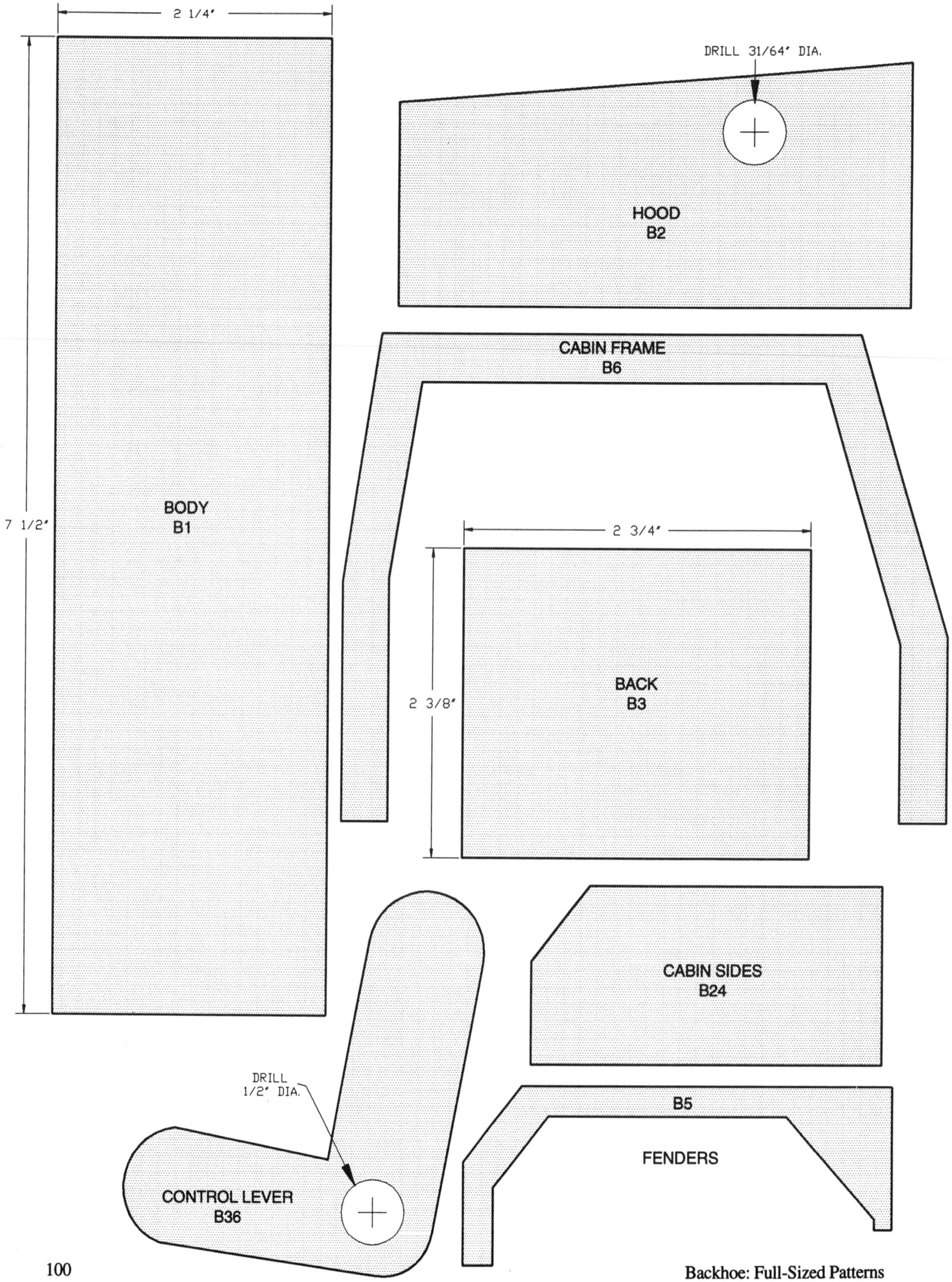

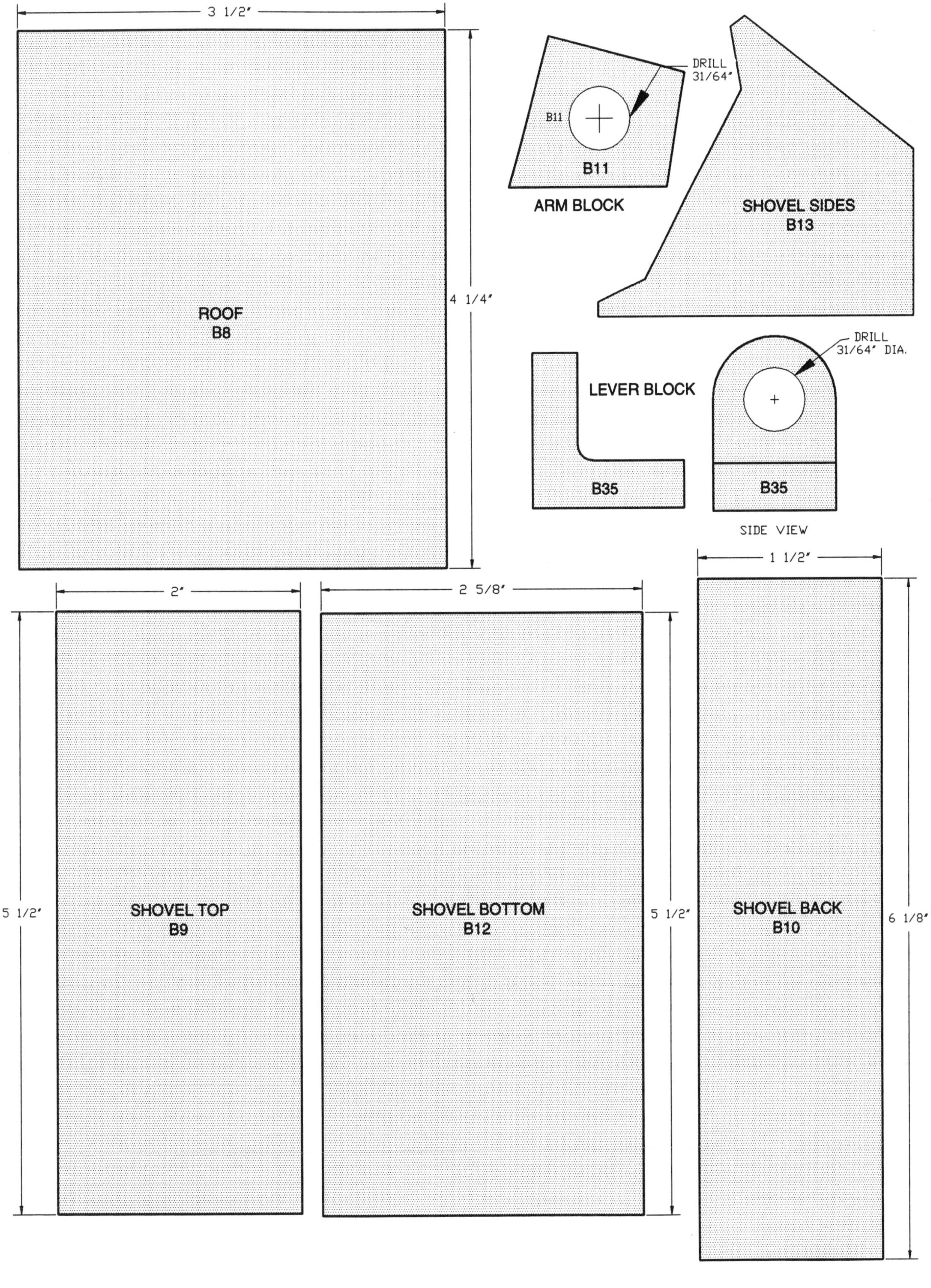
3 1/2"
4 1/4"
ROOF
B8
DRILL
31/64"
B11
B11
ARM BLOCK
SHOVEL SIDES
B13
LEVER BLOCK
DRILL
31/64" DIA.
B35
B35
SIDE VIEW
1 1/2"
2"
2 5/8"
5 1/2"
SHOVEL TOP
B9
SHOVEL BOTTOM
B12
5 1/2"
SHOVEL BACK
B10
6 1/8"

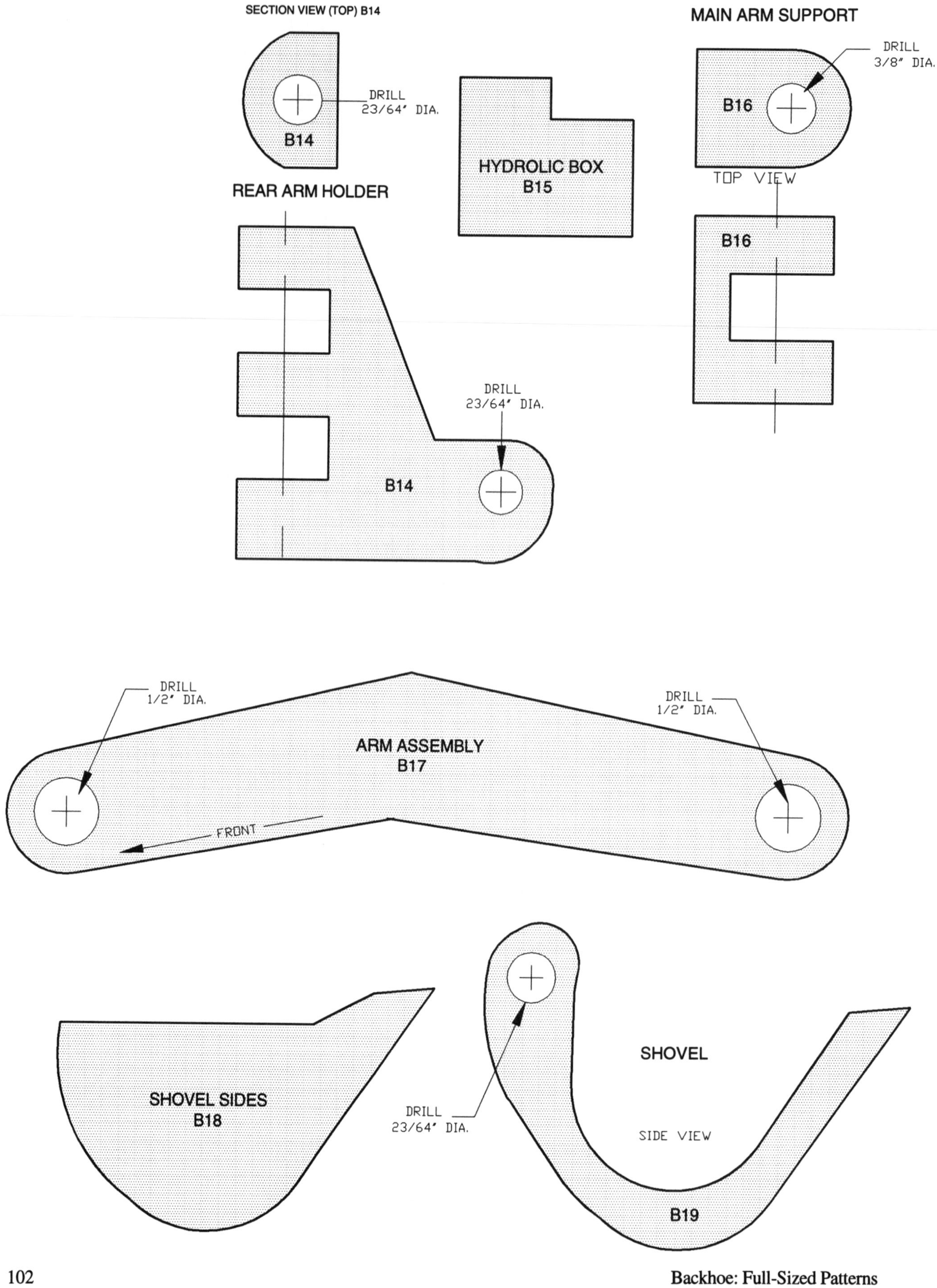
SECTION VIEW (TOP) B14
DRILL
23/64" DIA.
B14
HYDROLIC BOX
B15
MAIN ARM SUPPORT
DRILL
3/8" DIA.
B16
TOP VIEW
B16
REAR ARM HOLDER
DRILL
23/64" DIA.
B14
DRILL
1/2" DIA.
DRILL
1/2" DIA.
ARM ASSEMBLY
B17
FRONT
SHOVEL SIDES
B18
SHOVEL
DRILL
23/64" DIA.
SIDE VIEW
B19

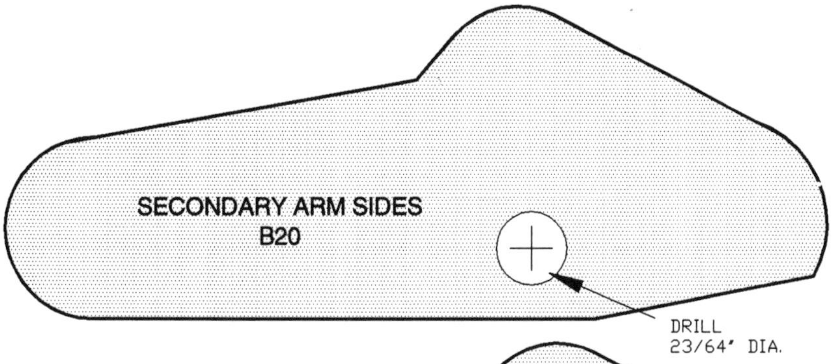
SECONDARY ARM SIDES
B20
DRILL
23/64" DIA.

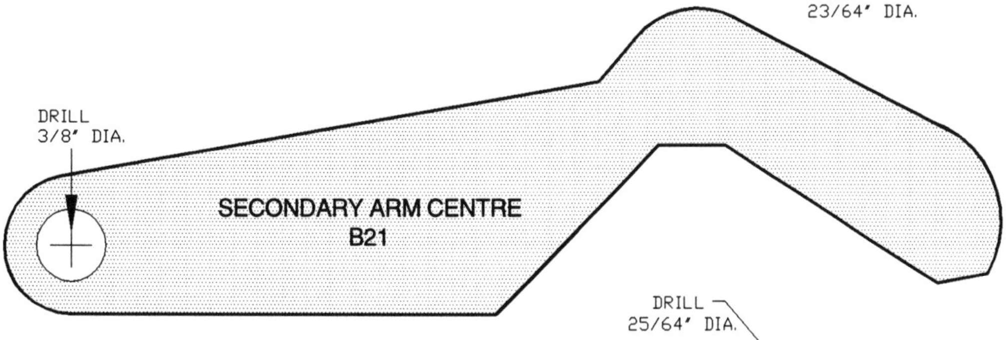
DRILL
3/8" DIA.
SECONDARY ARM CENTRE
B21
DRILL
25/64" DIA.

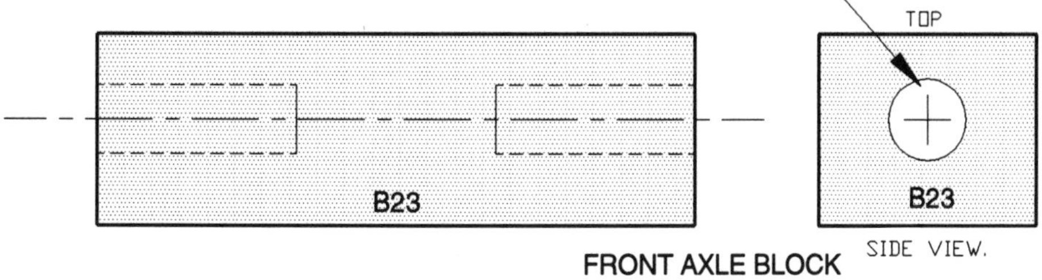
TOP
B23
B23
SIDE VIEW.

FRONT AXLE BLOCK

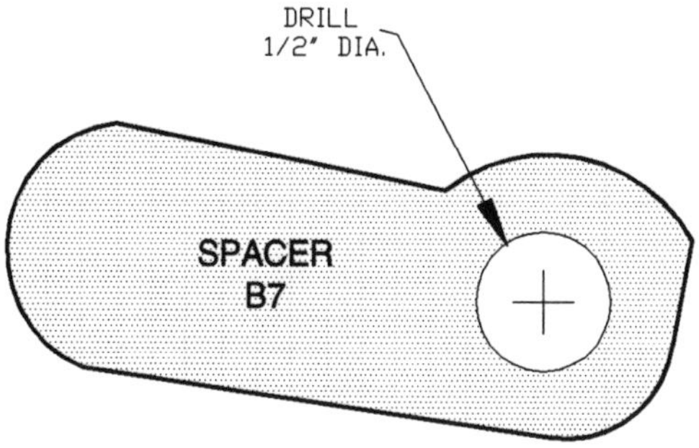
DRILL
1/2" DIA.
SPACER
B7

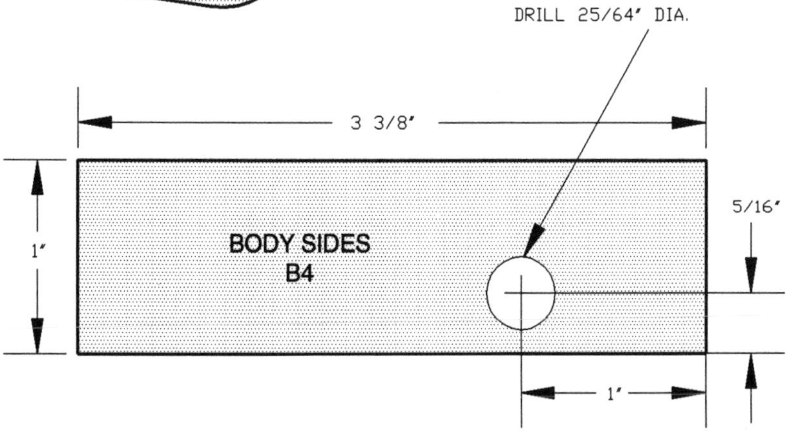
DRILL 25/64" DIA.
3 3/8"
1"
5/16"
BODY SIDES
B4
1"

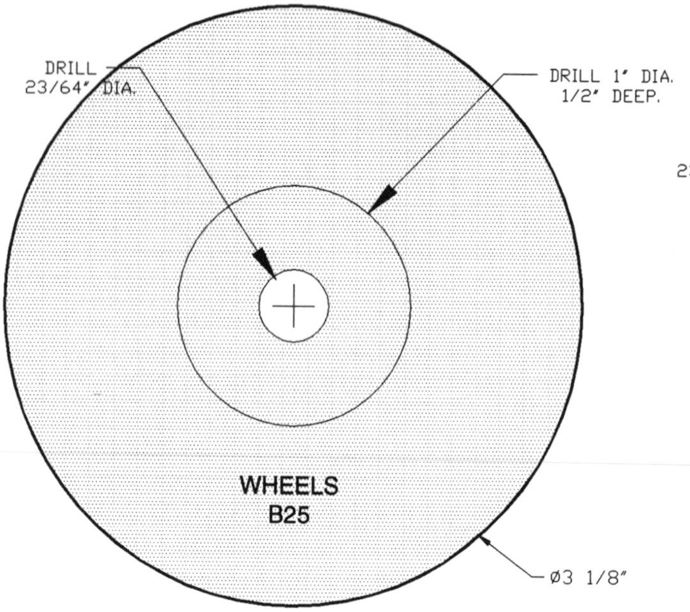
DRILL
23/64" DIA.
DRILL 1" DIA.
1/2" DEEP.
WHEELS
B25
Ø3 1/8"

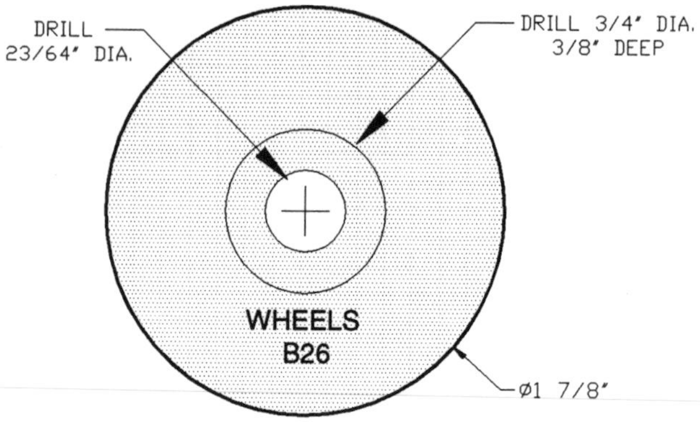
DRILL
23/64" DIA.
DRILL 3/4" DIA.
3/8" DEEP
WHEELS
B26
Ø1 7/8"

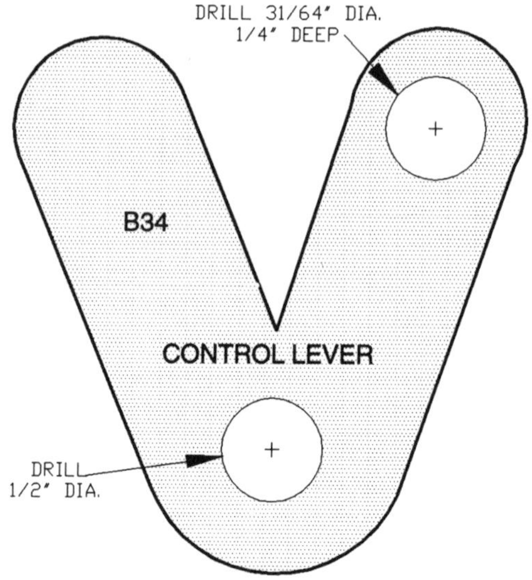
DRILL 31/64" DIA.
1/4" DEEP
B34
CONTROL LEVER
DRILL
1/2" DIA.

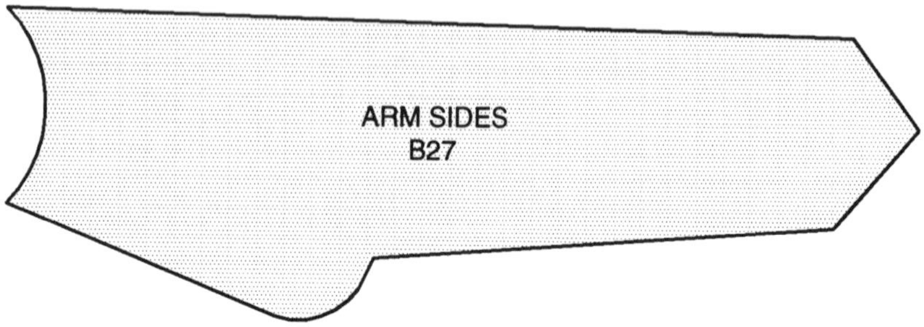
ARM SIDES
B27

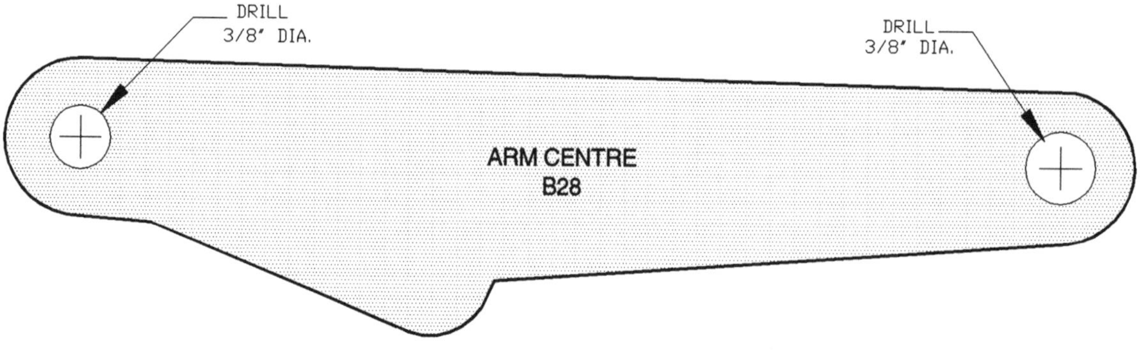
DRILL
3/8" DIA.
DRILL
3/8" DIA.
ARM CENTRE
B28

DRILL
1/2" DIA.
DRILL
1/2" DIA.
LEVER
B33

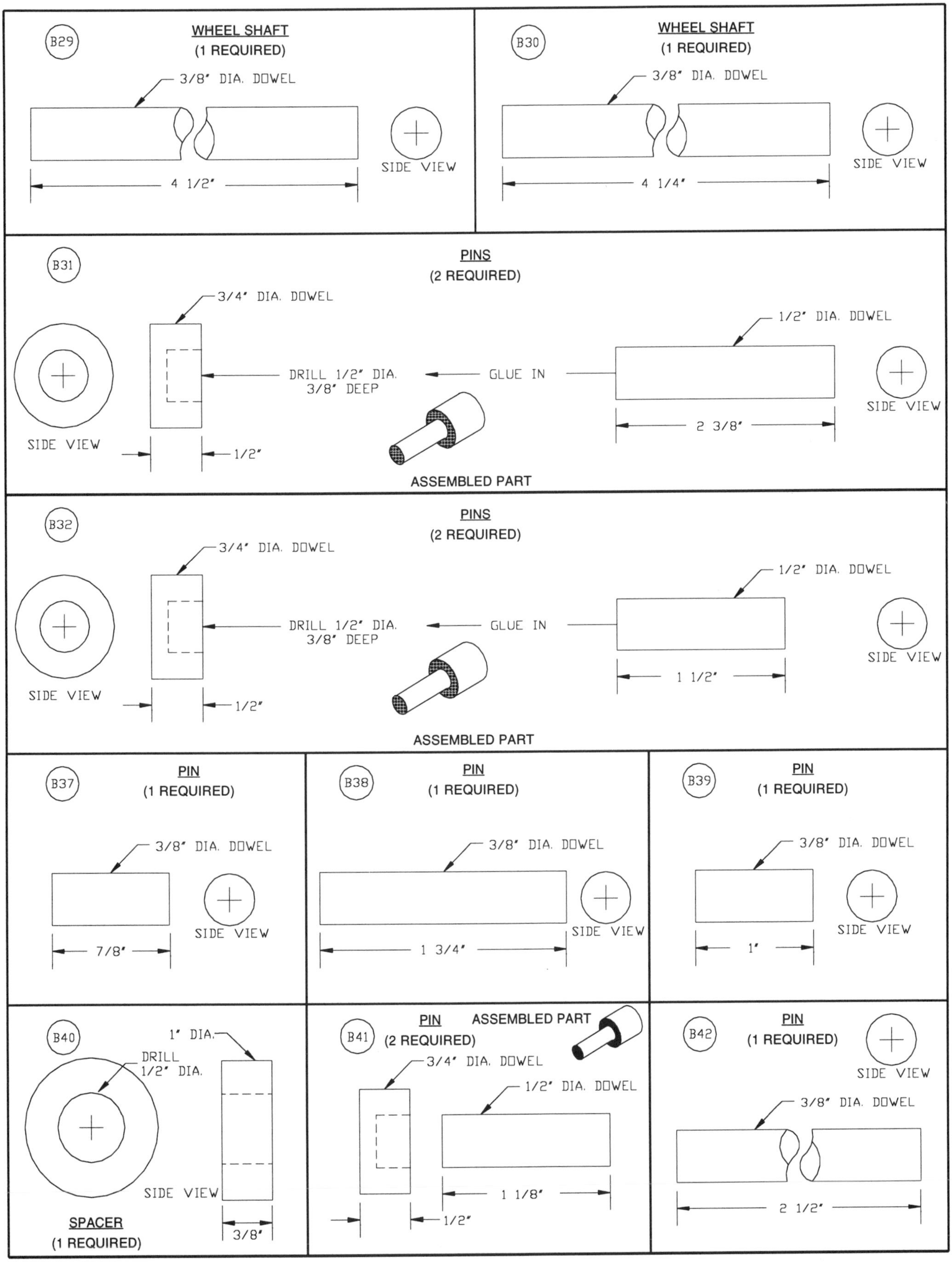
B29
WHEEL SHAFT
(1 REQUIRED)
3/8" DIA. DOWEL
4 1/2"
SIDE VIEW
B30
WHEEL SHAFT
(1 REQUIRED)
3/8" DIA. DOWEL
4 1/4"
SIDE VIEW
B31
PINS
(2 REQUIRED)
3/4" DIA. DOWEL
DRILL 1/2" DIA.
3/8" DEEP
GLUE IN
1/2" DIA. DOWEL
SIDE VIEW
1/2"
2 3/8"
SIDE VIEW
ASSEMBLED PART
B32
PINS
(2 REQUIRED)
3/4" DIA. DOWEL
DRILL 1/2" DIA.
3/8" DEEP
GLUE IN
1/2" DIA. DOWEL
SIDE VIEW
1/2"
1 1/2"
SIDE VIEW
ASSEMBLED PART
B37
PIN
(1 REQUIRED)
3/8" DIA. DOWEL
SIDE VIEW
7/8"
B38
PIN
(1 REQUIRED)
3/8" DIA. DOWEL
SIDE VIEW
1 3/4"
B39
PIN
(1 REQUIRED)
3/8" DIA. DOWEL
SIDE VIEW
1"
B40
1" DIA.
DRILL
1/2" DIA.
SIDE VIEW
3/8"
SPACER
(1 REQUIRED)
B41
PIN
(2 REQUIRED)
ASSEMBLED PART
3/4" DIA. DOWEL
1/2" DIA. DOWEL
1 1/8"
1/2"
B42
PIN
(1 REQUIRED)
SIDE VIEW
3/8" DIA. DOWEL
2 1/2"

Additional Information - Backhoe

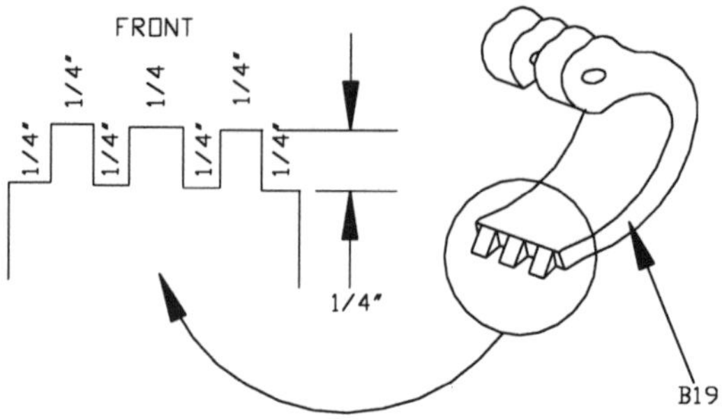

Using your Scroll Saw, cut grooves in part B19, as shown.

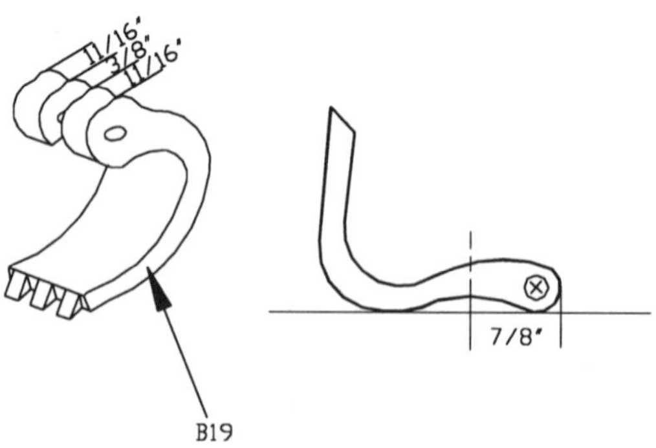

Using your Scroll Saw, cut groove in part B19.

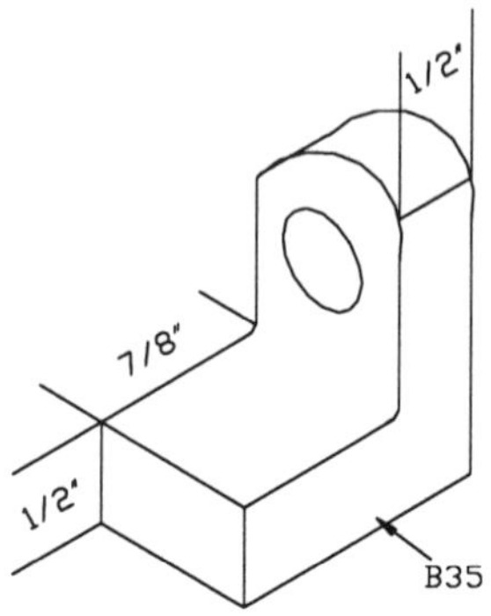

Drawing showing part B35 completed.

Important: When you cut this groove, make sure that arm assembly fits in tight. (this way, arm assembly stays up)

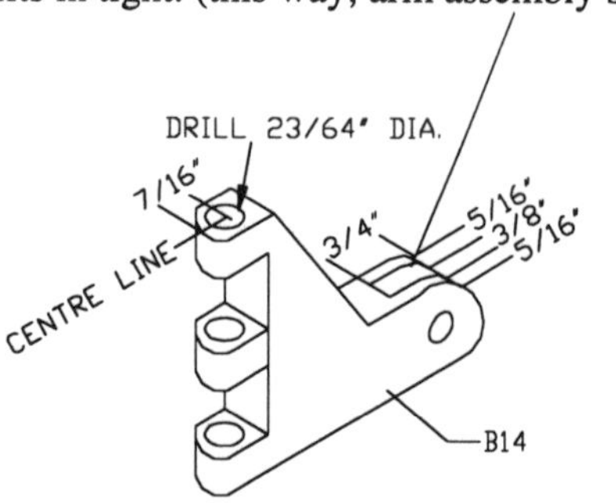

Using your Scroll Saw, cut groove in part B14, as shown.

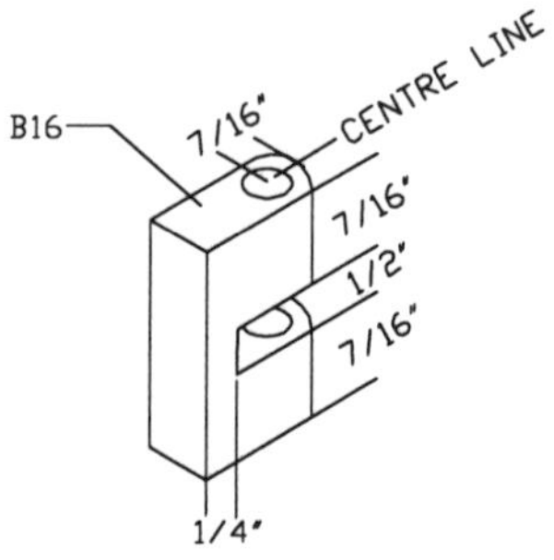

Using your Scroll Saw, cut groove in part B16, as shown.

Backhoe - Assembly Drawings

1

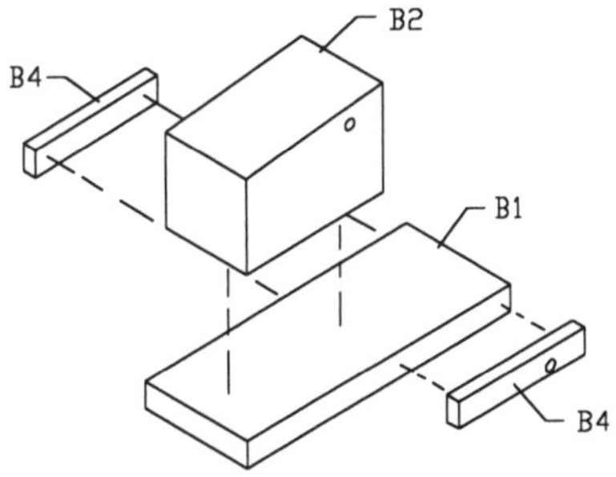

Glue hood B2 onto body B1.
Glue body sides B4.

2

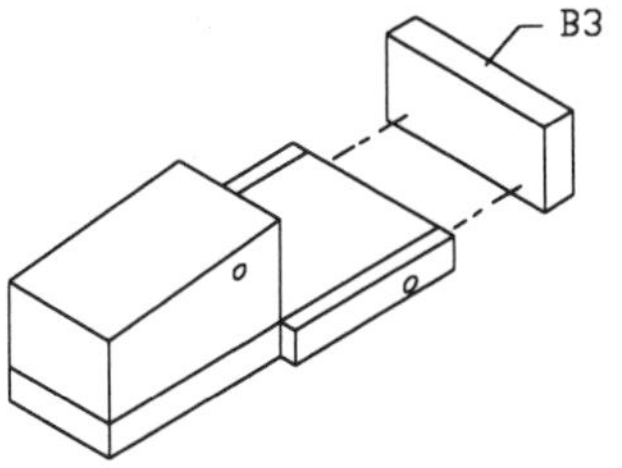

Glue back B3.

3

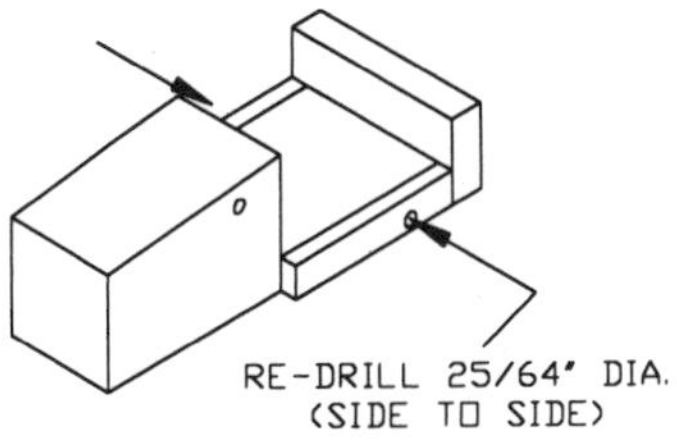

Re-drill hole through body.

4

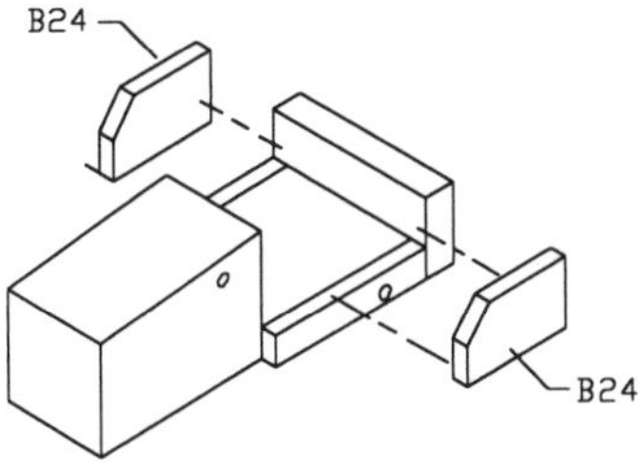

Glue cabin sides B24.

5

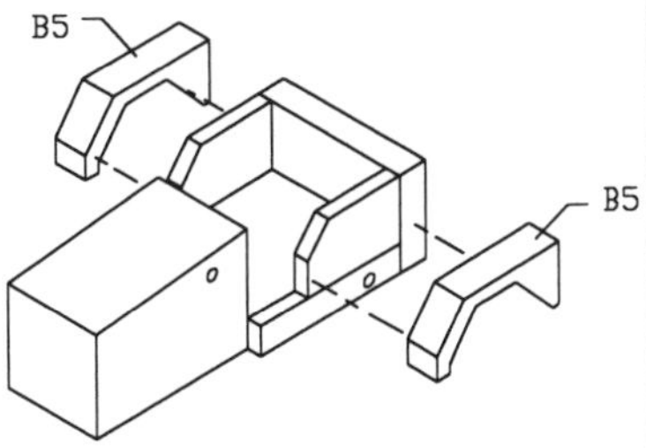

Glue fenders B5.

6

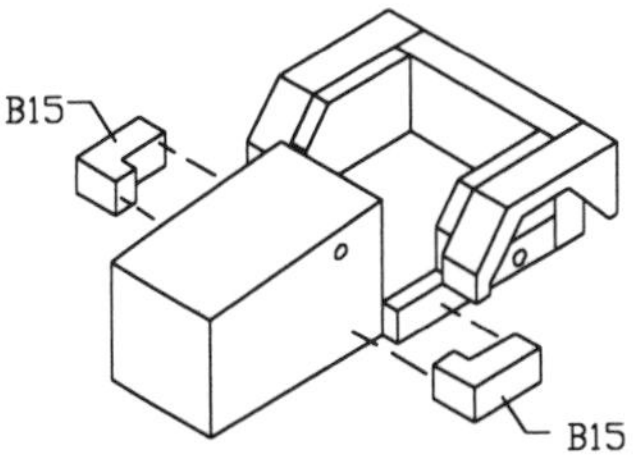

Glue hydrolic boxes B15.

7

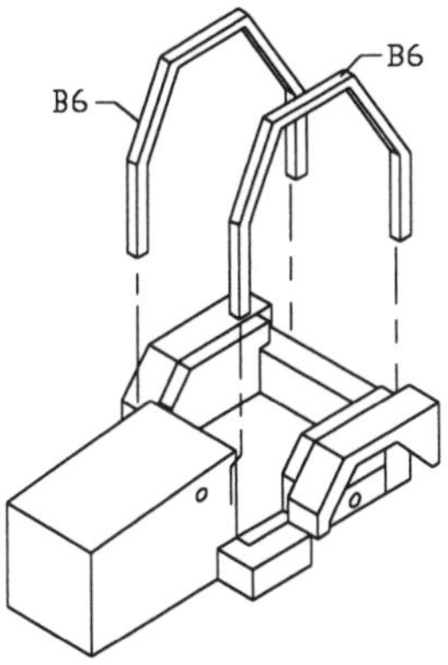

Glue cabin frames B6.

8

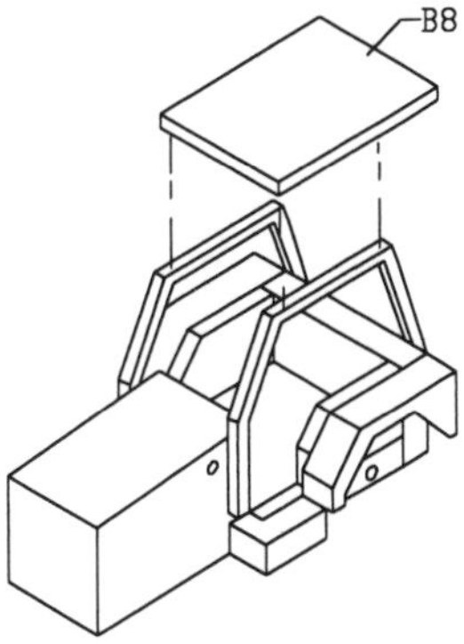

Glue roof B8.

9

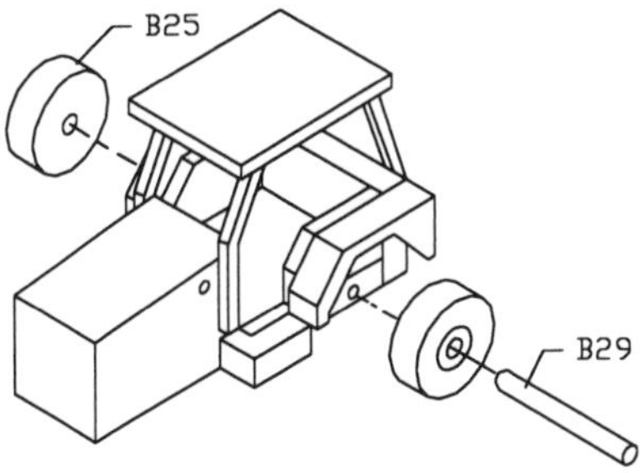

Insert wheel shaft B29 through wheels B25 and body.

10

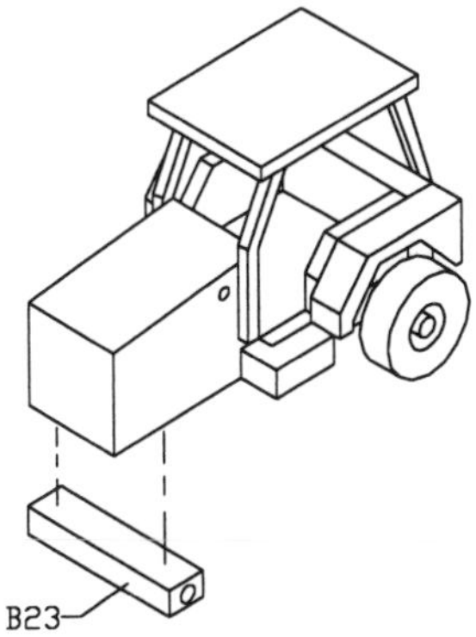

Glue front axle block B23.

11

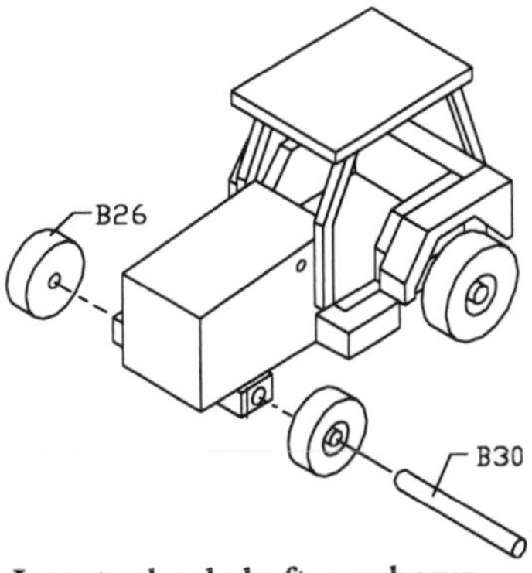

Insert wheel shaft, as shown.

12

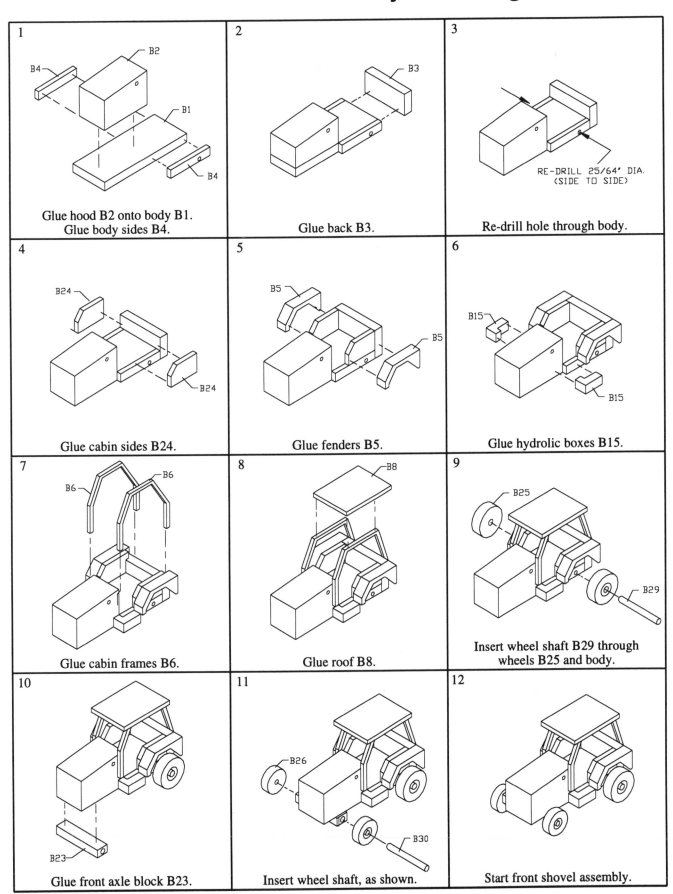

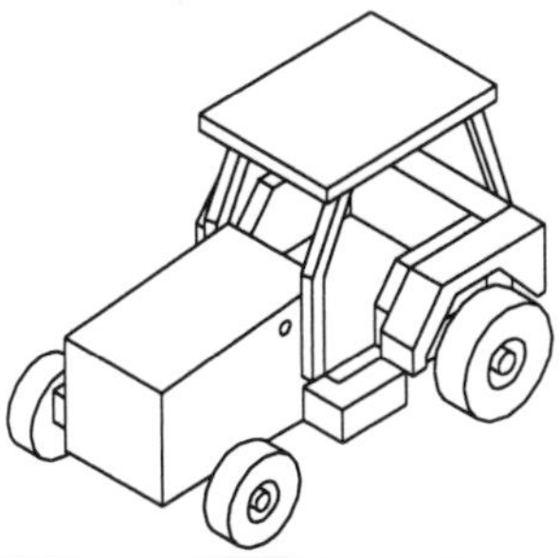

Start front shovel assembly.

13

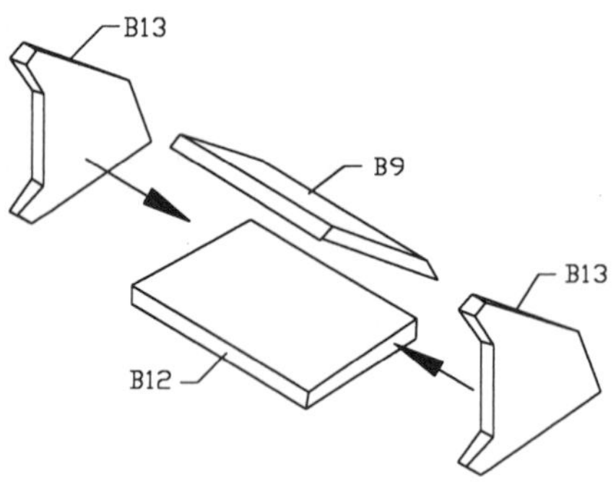

Glue shovel sides B13 onto shovel top B9 and bottom B12.

14

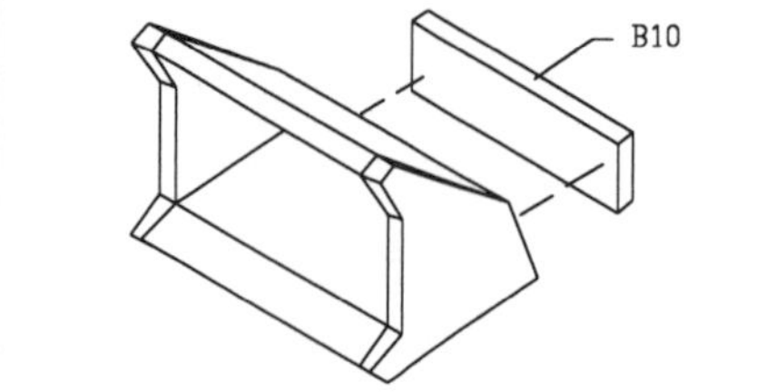

Glue shovel back B10.
See page 6 for hints on making shovel.

15

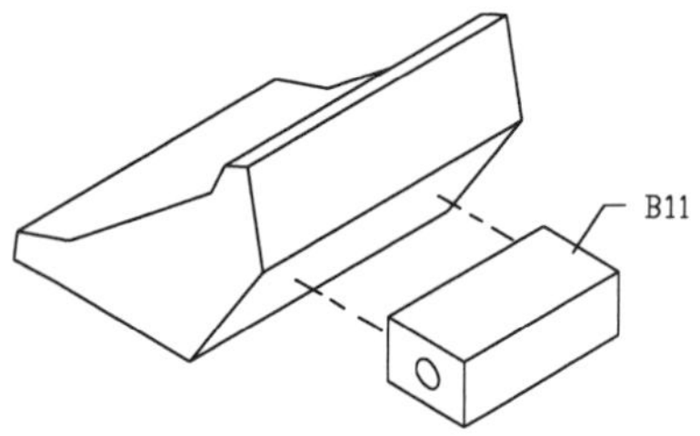

Glue arm block B11.

16

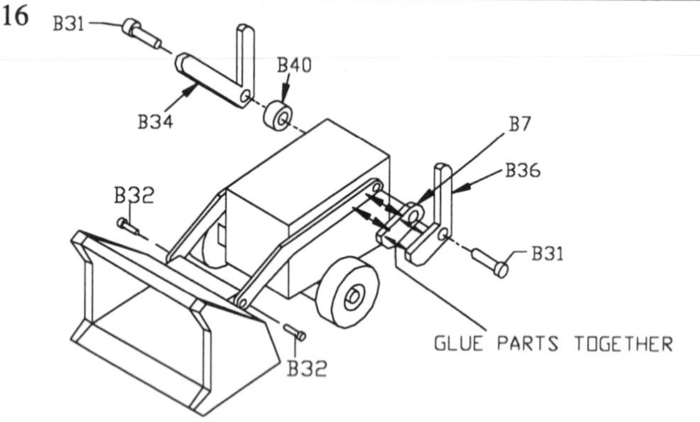

Assemble parts, as shown.

17

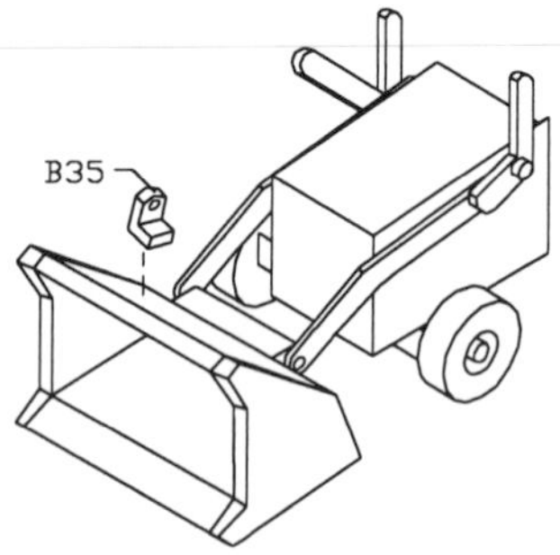

Glue lever block B35.

18

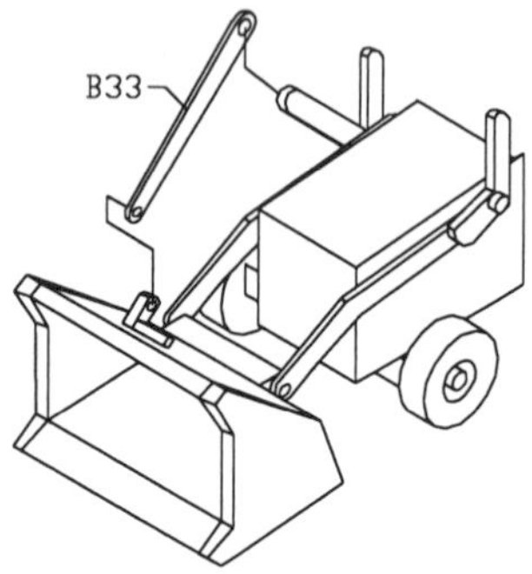

Install lever arm B33.

19

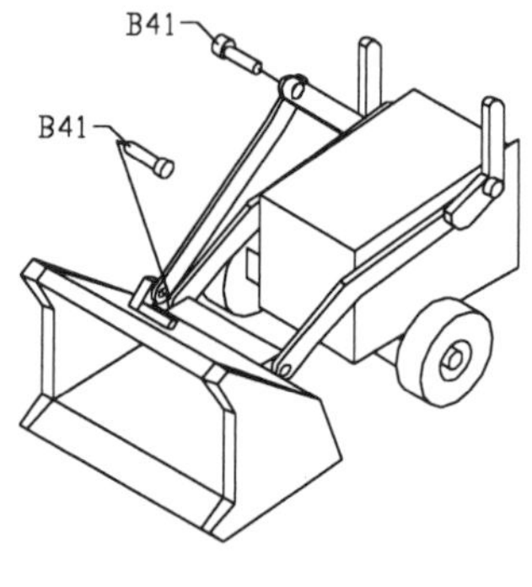

Insert locking pins B41.

20

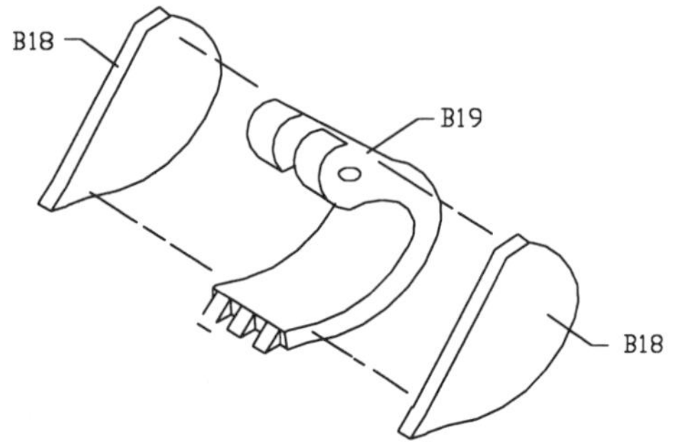

Glue shovel sides B18 onto shovel B19.

21

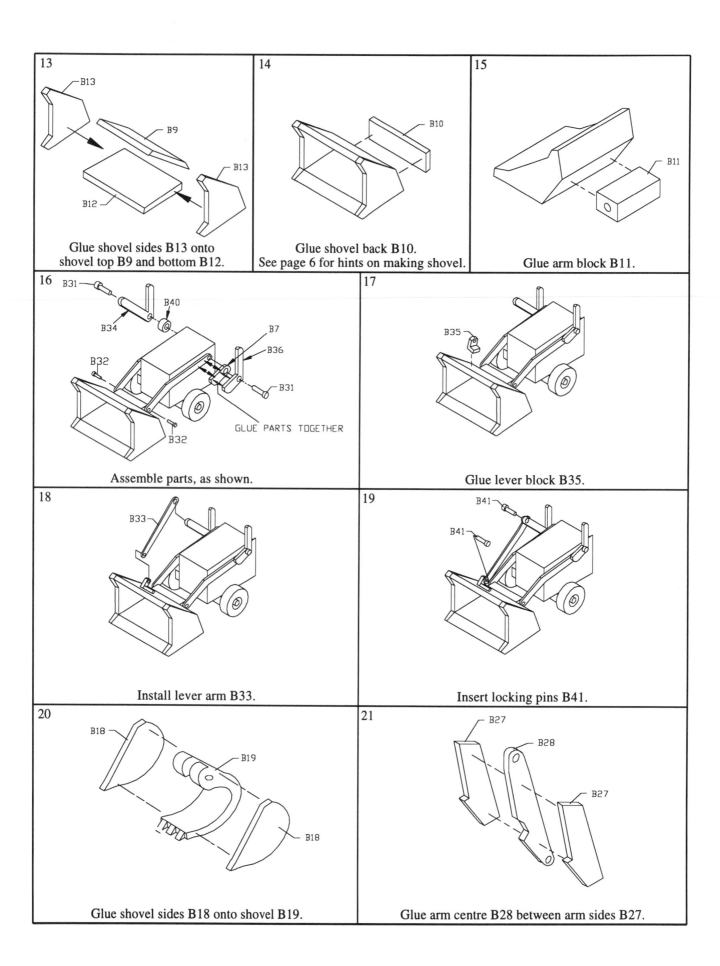

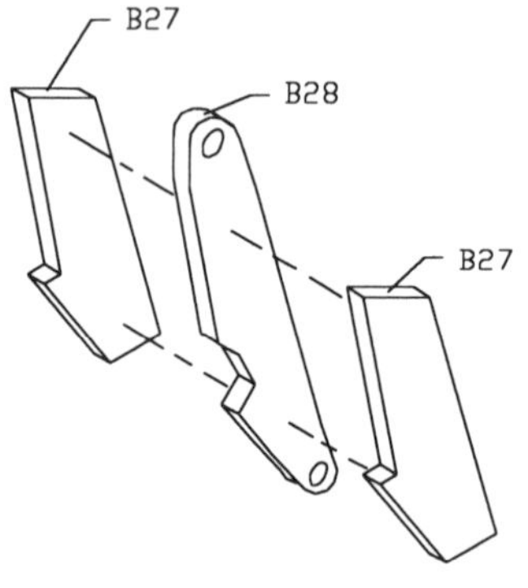

Glue arm centre B28 between arm sides B27.

22

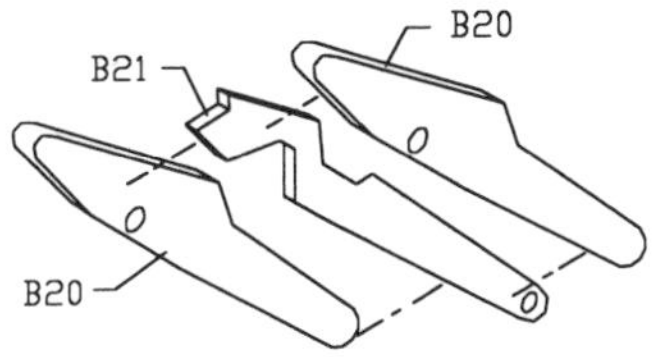

Glue secondary arm centre B21 between secondary arm sides B20.

23

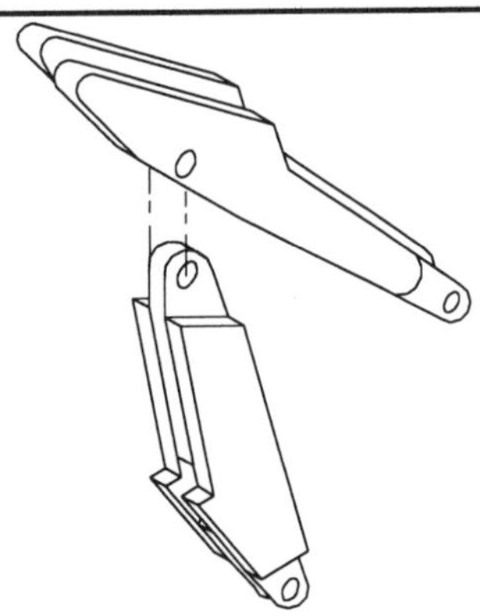

Slide arm assembly, as shown.

24

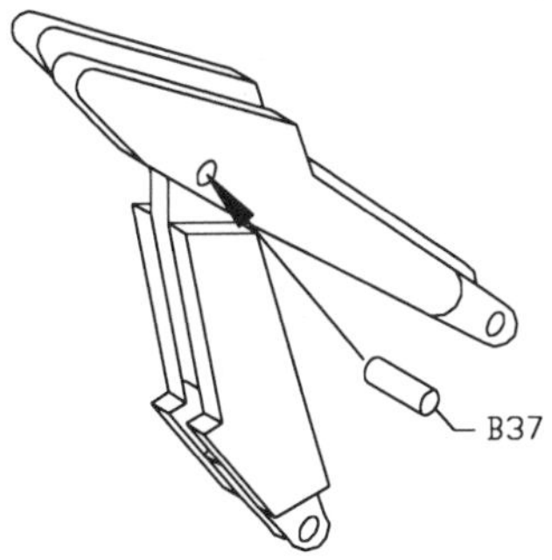

Insert pin B37.

25

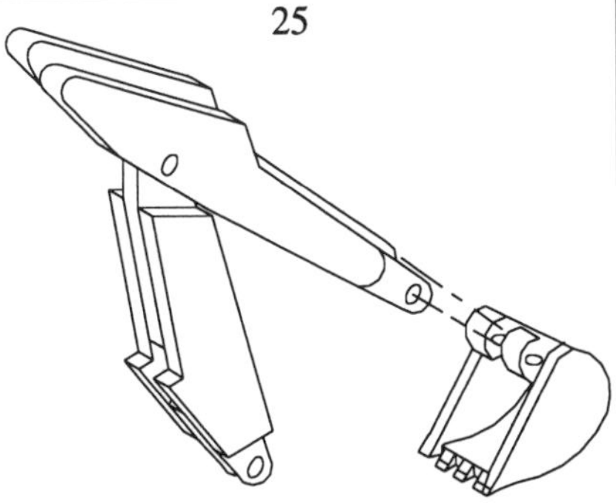

Slide shovel onto arm assembly.

26

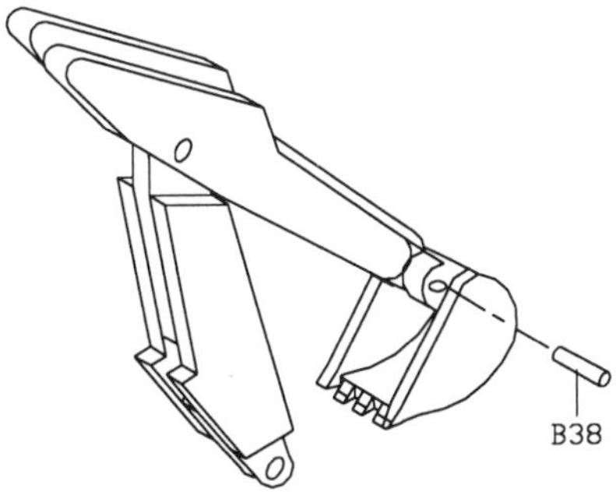

Insert pin B38.

27

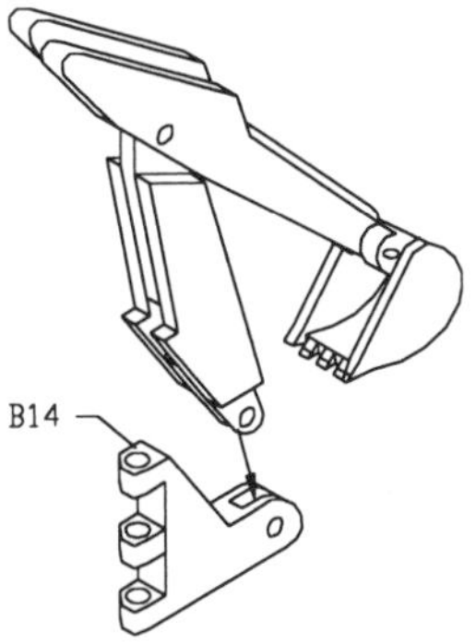

Slide arm assembly into holder B14.

28

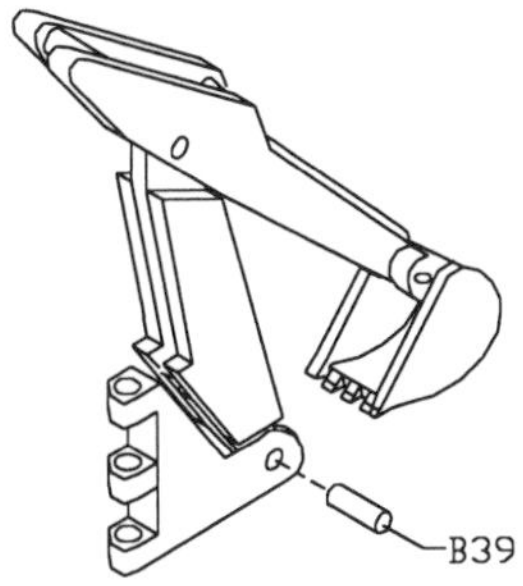

Insert pin B39.

29

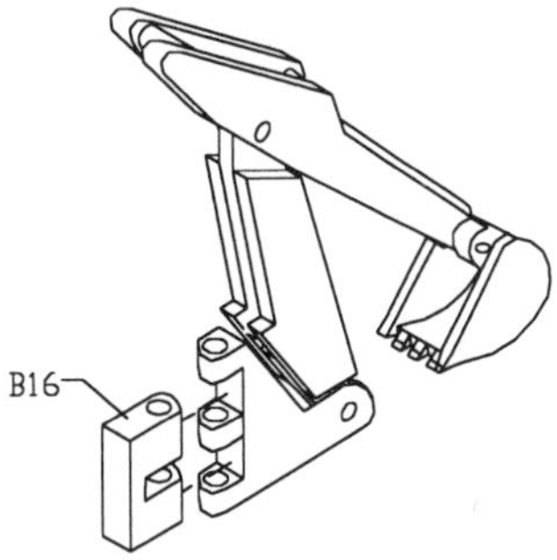

Slide arm assembly into support B16.

30

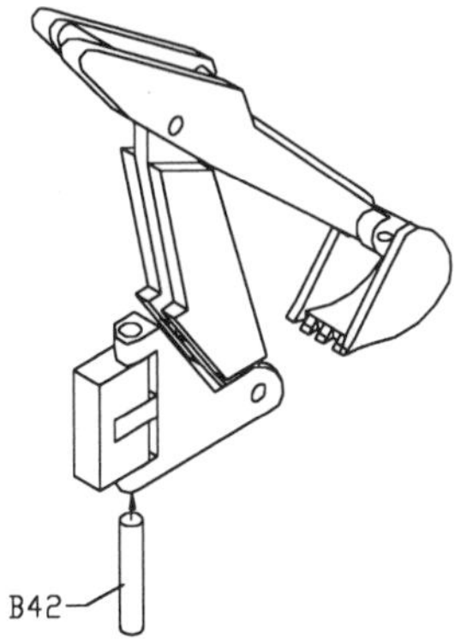

Insert pin B42.

31

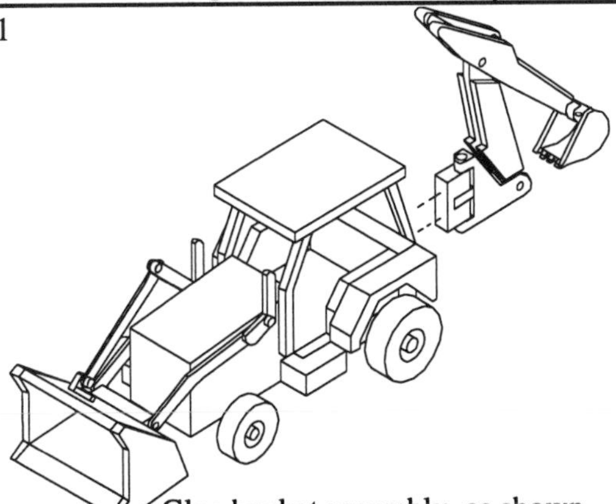

Glue bucket assembly, as shown.

32

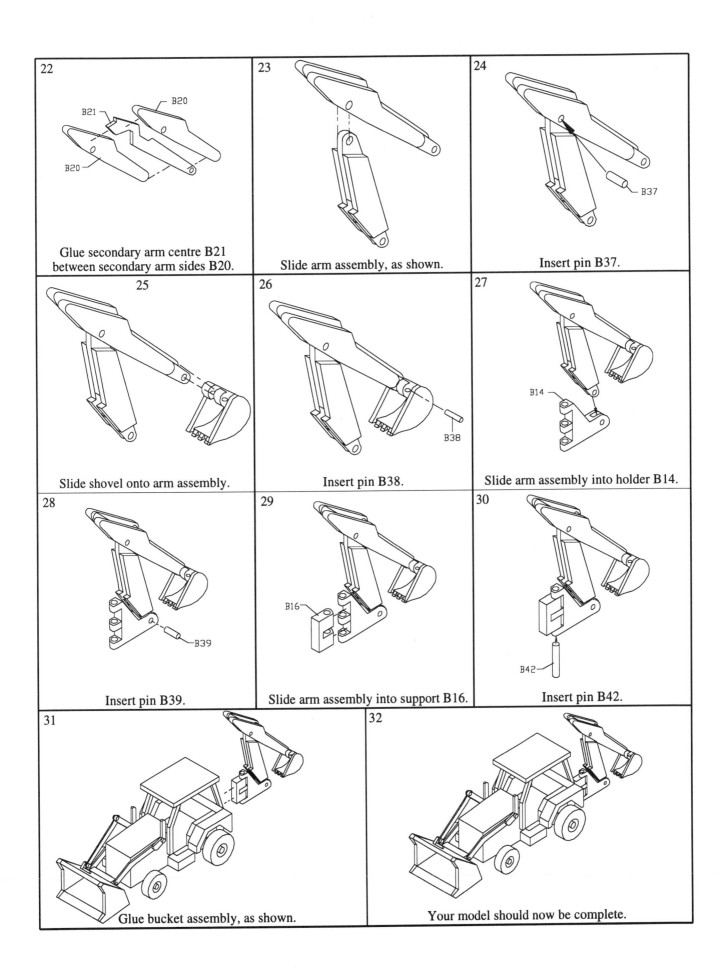

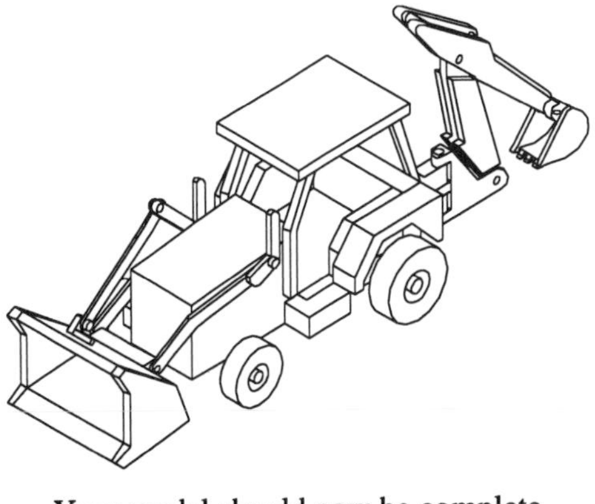

Your model should now be complete.

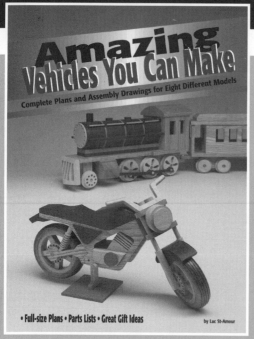

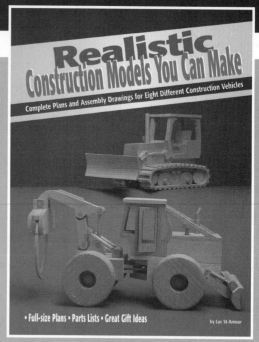

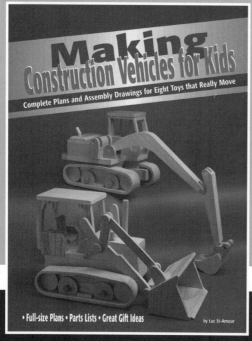

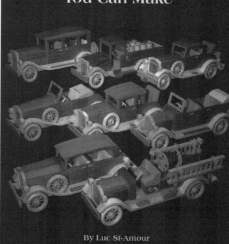